MECHANISMS OF EUKARYOTIC DNA RECOMBINATION

MECHANISMS OF EUKARYOTIC DNA RECOMBINATION

Edited by
MAX E. GOTTESMAN
HENRY J. VOGEL
College of Physicians and Surgeons
Columbia University
New York, New York

ACADEMIC PRESS, INC.
Harcourt Brace Jovanovich, Publishers
San Diego New York Boston
London Sydney Tokyo Toronto

This book is printed on acid-free paper. ∞

Academic Press, Inc.
1250 Sixth Avenue, San Diego, California 92101-4311

United Kingdom Edition published by
Academic Press Limited
24–28 Oval Road, London NW1 7DX

Library of Congress Cataloging-in-Publication Data

Mechanisms of eukaryotic DNA recombination / edited by Max E. Gottesman and Henry J. Vogel.
p. cm.
Includes bibliographical references and index.
ISBN 0-12-293445-8
1. Genetic recombination. 2. Recombinant DNA. I. Gottesman, Max E. II. Vogel, Henry J. (Henry James), date
[DNLM: 1. DNA Replication. 2. Gene Rearrangement. 3. Recombination, Genetic. QH 443 M486]
QH443.M44 1992
574.87'3282--dc20
DNLM/DLC
for Library of Congress 92-7144
CIP

PRINTED IN THE UNITED STATES OF AMERICA

92 93 94 95 96 97 QW 9 8 7 6 5 4 3 2 1

Contents

7 Sex-Dependent *de Novo* Methylation of the Transgene and Its Insertional Locus on Mouse Chromosome 13

X. WU, M. HADCHOUEL, H. FARZA, L. AMAR, AND C. POURCEL

8 A Role for Transcription in Antibody Switch Recombination

JOHANNA A. GRIFFIN AND MARY E. REABAN

PART III GENETIC RECOMBINATION IN *DROSOPHILA* AND *CAENORHABDITIS*

9 Genetic and Molecular Studies of a Simple Meiotic System

R. SCOTT HAWLEY AND ANNE E. ZITRON

10 Diversity among *Drosophila* Transposable Elements and in Their Effects on Gene Expression

STEPHEN M. MOUNT

Preface

Genetic recombination is a major force generating diversity in living organisms. Insight into the mechanism of genetic recombination has led to the ability to manipulate the genomes of prokaryotes and eukaryotes, including mammalian genomes, and to an initial understanding of biochemical processes in genetic recombination. The ability to perform targeted DNA integration into the genomes of different eukaryotes has facilitated the analysis of numerous cellular processes and the construction of disease models. In this context, this volume presents recent advances in eukaryotic genetic recombination, addressing targeted DNA integration in mammalian genomes, the manipulation of the mouse genome, genetic recombination in *Drosophila*, *Caenorhabditis*, and yeast, and genome organization and genetic recombination in protozoa.

It is a pleasure to acknowledge the advice and help of Dr. Paul Berg, Dr. Franklin D. Costantini, Dr. Robin Holliday, Dr. Stephen M. Mount, Dr. Rodney J. Rothstein, Dr. Maxine F. Singer, and Dr. Lex H. T. Van der Ploeg.

We are grateful for the continued interest of Dr. Donald F. Tapley and for the fine support of the College of Physicians and Surgeons (P&S) of Columbia University without which this volume would not have reached fruition. This volume was developed from a P&S Biomedical Sciences Symposium held at Arden House, on the Harriman Campus of Columbia University.

Max E. Gottesman
Henry J. Vogel

PART I

TARGETED DNA INTEGRATION IN MAMMALIAN CELLS

1

Chromatid Interactions during Intrachromosomal Recombination in Mammalian Cells

RONI J. BOLLAG AND R. MICHAEL LISKAY
Departments of Human Genetics and Therapeutic Radiology
Yale University School of Medicine
New Haven, Connecticut 06510

INTRODUCTION

Intrachromosomal recombination can involve two discrete types of chromatid interactions: intrachromatid, which occurs between linked sequences on a DNA molecule and sister chromatid, which occurs between equally or unequally paired homologous sequences following DNA replication (see Fig. 1). In most cell systems, equal sister chromatid exchange has no genetic consequence.

Gene conversion is the predominant mode of intrachromosomal recombination in mouse L cells (1). Since such nonreciprocal exchanges involve unidirectional information transfer between DNA molecules, sister chromatid and intrachromatid convertants appear identical and do not impart information concerning the interaction involved. On the other hand, reciprocal exchanges often lead to distinct products in which the type of interaction can be ascertained. For instance, intrachromatid reciprocal exchange between genes in an inverted orientation generates inversion products, whereas sister chromatid exchanges lead to inviable aberrant products that are not recovered (2). Simple intrachromatid reciprocal exchange between direct repeats should generate two products: a circular DNA molecule with one gene and a single hybrid gene located in the chromosome and

Mechanisms of Eukaryotic DNA Recombination

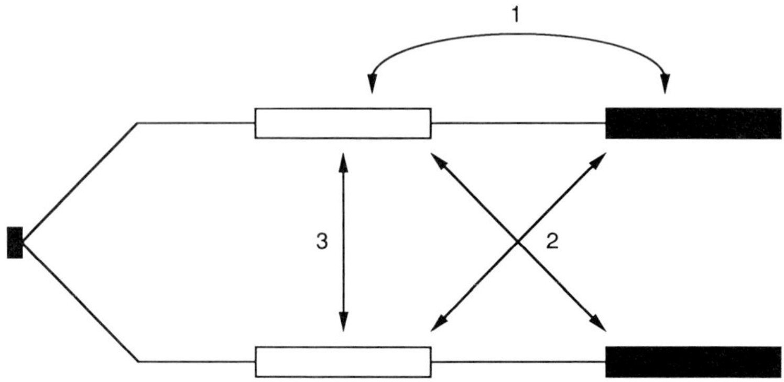

Fig. 1. Types of chromatid interactions. Diagrammed are sister chromatids with pairs of repeated genes (blocks). (1) Intrachromatid exchange between linked genes on one chromatid. Interactions of this type can occur at the two-strand or at the four-strand stage. (2) Unequal sister chromatid exchange between genes at nonallelic positions on sister chromatids. Such interactions require unequal pairing. (3) Equal sister chromatid exchange between identical sequences at allelic positions. Such interactions, if they occur, have no genetic consequence in recombination studies.

lacking the sequence between the two interacting sequences. Sister chromatid reciprocal exchange between a pair of direct repeats gives rise to one chromatid with a triplication and one chromatid with only a single gene and deletion of the intervening sequence. During cell propagation without selection, a circular DNA molecule should be lost. Chromatids with only a single gene are designated *deletion products* in this study. Deletion products of intrachromatid exchange are indistinguishable from deletion products of sister chromatid exchange. In the *tk*-selective system used in the present study, the reciprocal product that would harbor a wild-type gene would depend on the orientations of mutations within the genes (see Figs. 2 and 3). The outcomes of analyses of these products form the basis of this study.

Orientation I refers to a configuration in which mutations are proximal to the intervening sequences, whereas the orientation II construct harbors genes with mutations distal to the intervening sequences. Reciprocal recombination between direct repeats in orientation I generates deletion products, whether the process involves sister chromatid or intrachromatid exchange (Fig. 2). By contrast, sister chromatid reciprocal exchange results primarily in triplication recombinants between orientation II direct repeats, whereas intrachromatid exchange is generally unproductive (Fig. 3). By comparing the observed proportions of reciprocal exchanges obtained in either orientation, we have estimated the relative frequencies of sister chromatid and intrachromatid reciprocal exchanges. Furthermore, an analysis of recombi-

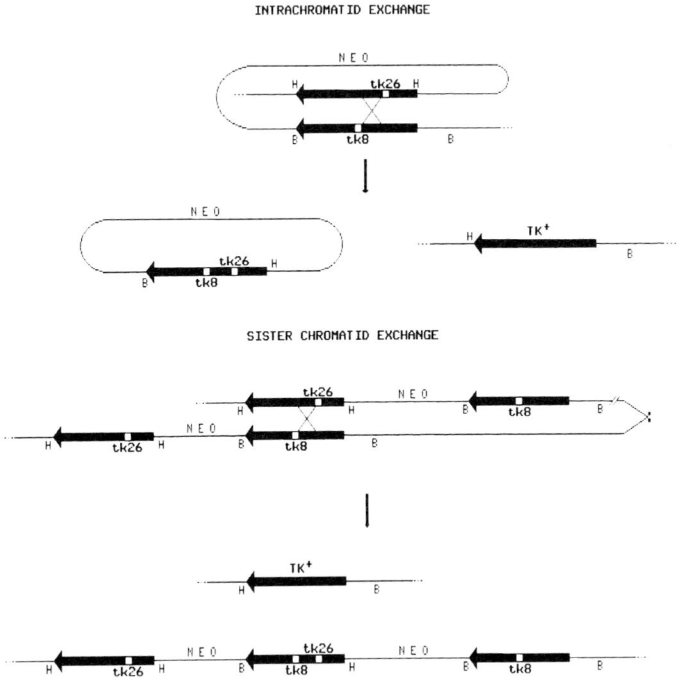

Fig. 2. Simple reciprocal recombinations between direct repeats in orientation I. Mutations are proximal to intervening sequence. Open blocks designate *Xho I* linker insertion mutations *tk26* or *tk8*. Top panel depicts intrachromatid exchange. H refers to *Hin*dIII restriction sites flanking the *tk26* allele. B refers to *Bam*HI restriction sites flanking *tk8* allele. Simple reciprocal exchange in the region between the mutations leads to a wild-type sequence in the chromosome with looping out of the double mutant gene and *neo*. Reciprocal products have hybrid flanking markers, H and B. Bottom panel depicts sister chromatid exchange. Centromere location is hypothetical; the relative orientation of the centromere is inconsequential. Simple reciprocal exchange between *tk8* and *tk26* alleles on either strand generates a single wild-type gene on one chromatid, and a double-mutant gene flanked by two mutant genes on the sister chromatid, which is lost. Deletion products of either intrachromatid or sister chromatid exchange are indistinguishable and both render the cell phenotypically G418^{s}.

nants between direct repeats has provided evidence for products of reciprocal exchange that are associated with a separated conversion tract.

RESULTS AND DISCUSSION

Intrachromosomal recombination has been studied most extensively with *tk* gene constructs and, to a lesser extent, using *neo* genes. Initial studies with the *tk* system used in the present study suggested that the predominant

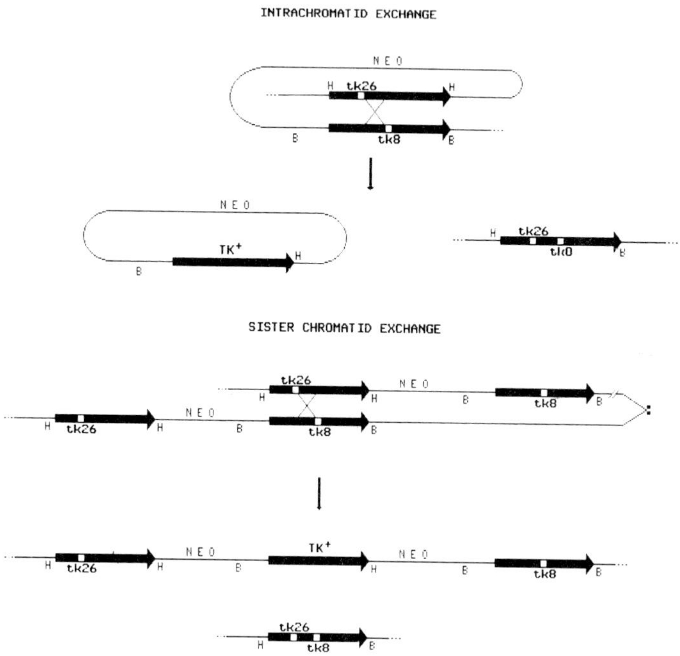

Fig. 3. Simple reciprocal recombinations between direct repeats in orientation II. Mutations are distal to intervening sequence on both genes. Top panel depicts intrachromatid exchange leading to looping out of the wild-type gene and generation of a double-mutant gene in the chromosome. The circular pop-out is not recovered; intrachromatid reciprocal exchange is not productive except in the special case discussed in Fig. 4. Bottom panel depicts sister chromatid reciprocal exchange resulting in a wild-type gene on the sister chromatid flanked by two mutant genes, resulting in a triplication product.

mode of recombination was nonreciprocal in nature, but that reciprocal exchanges composed 15–20% of all productive recombinations (1). Experiments with the *tk* genes in mouse L cells provided evidence that recombination rates were intrinsic to a given construct and not subject to profound influences of genomic integration sites, which presumably vary from clonal cell line to cell line (3). Rates of recombination, derived by fluctuation analyses, were first determined for recombination substrates arranged as full-length direct repeats in orientation I (4). These lines were generated by introduction of the constructs, using the calcium phosphate coprecipitation procedure. The rates from these previously reported studies are reproduced in Table I.

In the present study, the directions of both *tk* genes were reversed on the recombination substrate to produce direct repeats in orientation II, and the construct was introduced into mouse L cells by direct nuclear microinjec-

TABLE I
Rates of Recombination between Direct Repeats

Line[a]	Rate[b]
Orientation I	
pJS-3-10	3.5×10^{-6}
pJS-3-3-3	4.8×10^{-6}
pJS-3-3-4	2.6×10^{-6}
Orientation II	
pRB-1-3a-1	1.3×10^{-6}
pRB-1-C-1	1.9×10^{-6}
pRB-1-l-1-1	4.0×10^{-6}
pRB-1-E-3-1	1.4×10^{-6}
pRB-1-5-1-2	0.4×10^{-6}

[a]All lines harbor a single pair of *tk* alleles stably integrated into the genome.

[b]Rates reflect recombinations/cell/generation.

tion. As in the earlier studies, only cell lines with a single copy of the recombination construct were chosen for further analysis. Rates were determined for these lines in a fashion similar to that previously reported (1–4) and are recorded in Table I. Overall rates of recombination in orientation I and II are not dramatically different, suggesting that a comparison of recombination properties in the two orientations is valid.

The focus of this study was a comparison of products of reciprocal recombination from orientation I and orientation II direct repeats. Initial characterization of recombinants relied on genetic analysis to determine G418 resistance (G418^r) or sensitivity (G418^s), denoting the presence or absence of the *neo* gene. As noted in Figs. 2 and 3, G418^s products most likely represent deletion products of reciprocal exchange. For orientation I, G418^r products reflect almost exclusively gene conversions, although, as mentioned subsequently, a small minority may represent triplications resulting from sister chromatid reciprocal exchange accompanied by separated gene conversion. Earlier studies presented only a limited analysis of products from orientation I (1,4), and this analysis has been expanded in the present study. A cumulative tabulation of recombinants determined by G418 testing is provided in Table II. As shown in Table II, the proportion of conversions is 83.5% (or slightly below this figure when account is made of triplication products).

The analysis of orientation II recombinants relied primarily on molecular analysis of independent hypoxanthine, aminopterin, thymidine-resistant

TABLE II
Types of Recombinants

Type	Orientation I	Orientation II
Conversions	81 (83.5%)	117 (81.3%)
Triplications	N.A.[a]	22 (15.3%)
Deletions	16 (16.5%)	5 (3.4%)

[a]Not available. Numbers for orientation I represent results of a genetic analysis, whereby conversions are considered to be G418^r, and deletions are considered to be G418^s. Triplications, if they occur, would be detected as G418^r and classified as conversions.

(HAT^r) colonies by Southern blot hybridization. Nevertheless, a genetic test was performed, and the proportion of G418^s products was slightly greater than 3% of the total. The significance of these products will be addressed. The molecular hybridization analysis was performed to distinguish, among orientation II G418^r recombinants, sister chromatid reciprocal exchanges resulting in triplication (see Fig. 3) from gene conversions of either allele *tk8* or *tk26*. The result of this analysis, including the contribution of G418^s recombinants determined genetically, is tabulated in Table II. Of note in this analysis, orientation II reciprocal exchanges to generate triplications, presumably arising through sister chromatid interactions, compose 15.3% of all recombinant products. Consistent with previous recombinant analyses in mouse L cells, gene conversions represent the majority of all recombinants (81.5%).

The primary goal of the present study is to infer the types of chromatid interactions between repeated genes. Because gene conversions are noninformative, the strategy involves a comparison of reciprocal exchanges between orientation I and orientation II direct repeats. Whereas both sister chromatid and intrachromatid interactions can produce deletion products, only sister chromatid interactions can give rise to triplication products in orientation II. Thus, the proportions of intrachromatid and sister chromatid exchanges can be ascertained by dissecting the proportions of overall reciprocal exchanges in orientation I into a sister chromatid component, calculated from orientation II triplications, and an intrachromatid component, which is inferred to be the remainder of that figure and to reflect intrachromatid exchanges that are not recovered in orientation II. Although we classify the deletion and triplication products observed in this study as reciprocal exchanges, there is the formal possibility that these products can be generated by sister chromatid conversion [also referred to as plasmid conversion (5) or unequal gene conversion (6)], as proposed for intrachromosomal recombination in yeast (5–8). In our system, sister chromatid conversion would require conversion of large heterologies [4.4 kilobases (kb)]. Since studies in mouse L cells suggest that conversion of heterologies larger than 1 kb occurs at a rate at least 20-fold less than that observed for recombination generating the products diagrammed in Figs. 2 and 3 (3), we favor reciprocal exchange, rather than conversion, as the primary mechanism for generating these products.

The calculation

Orientation I deletions = orientation II triplications + intrachromatid exchanges

presumes that the proportion of reciprocal exchanges in orientation I resulting in deletion reflect both sister chromatid and intrachromatid exchanges, whereas the proportion of triplication products in orientation II reflects only the sister chromatid component of the orientation I deletions. It should be noted that since orientation II intrachromatid exchanges are generally unproductive, the relative proportion of sister chromatid exchanges (determined by the number of triplications divided by total recombination products) is slightly overestimated, since the denominator is lower by the deficit of intrachromatid exchanges.

Nevertheless, the conclusion can be reached that in this system sister chromatid exchanges predominate. Since fully 15.3% of all orientation II products are sister chromatid reciprocal exchanges (triplications), whereas 16.5% of orientation I products include both intrachromatid and sister chromatid reciprocal exchanges, the majority of products from direct repeats (15.3 of 16.5, or greater than 90%) are sister chromatid rather than intrachromatid. This conclusion is consistent with earlier studies with inverted repeats in which intrachromatid inversions composed only 6% of the total products; for inverted repeat recombination, sister chromatid exchange is not productive (2).

The conclusion that sister chromatid exchange between linked repeats predominates in mammalian cells is interesting in light of a recent report that intrachromatid exchanges predominate in yeast. These studies in yeast involve recombination between orientation I and between orientation II direct repeats of *LEU2* or *HIS3*, analyzed with a rationale identical to that detailed here. Klein (6) found that mitotic reciprocal exchange resulting in gene deletion ("pop-out") for orientation I represented 50% of all events, whereas the remainder represented mitotic gene conversion. Conversely, triplications in orientation II comprised only between 0 and 7% of recombinants, depending on the alleles used. The proportions of pop-outs in these studies are greater than those reported previously in yeast for duplications of the *HIS4* marker (9). In this latter study, only 12% of orientation I products were pop-outs, whereas in a limited analysis, triplications were readily apparent in orientation II. These disparate results in yeast suggest that caution should be used in assessing the universality of our conclusion, since sister chromatid exchanges may not predominate using other marker systems or in other mammalian cell lines.

The observation that *tk* recombination in L cells more frequently involves sister chromatid than intrachromatid interactions suggests that the majority of reciprocal exchanges occur during or subsequent to DNA replication (S phase or G_2 phase). Indeed, as intrachromatid exchanges can occur following DNA replication as well as before, there is at present no evidence that reciprocal exchange occurs during G_1 phase. Roman and Fabre (10) have proposed that reciprocal exchanges in mitotic yeast can occur in G_2 phase, often preceded by gene conversion, which is thought to predominate in G_1 phase. This observation suggests that it may not be appropriate to extend the conclusion that sister chromatid interactions are more frequent among all events, but rather that the gene-conversion class of recombinants may be temporally distinct. An earlier study with inverted *tk* genes in L cells provided evidence that reciprocal exchanges can be associated with gene conversion, suggesting that the two processes are mechanistically related (2).

One class of event that deserves additional attention is the category of deletions in orientation II. As depicted in Fig. 3, simple reciprocal exchange is not predicted to produce deletions in orientation II, as the chromosomal product would harbor a double-mutant gene. Reciprocal exchange accompanied by adjacent conversion similarly is inconsistent with deletion formation, as the product would still retain one mutation. The most likely mechanism for generating deletion recombinants involves reciprocal exchange accompanied by a separated conversion tract. Reciprocal exchange to one side of the gene, associated with active conversion of the distal mutation, should generate the observed product, as diagrammed in Fig. 4.

The predicted frequency of independent reciprocal and gene conversion events cannot account for the observed products. Therefore, the high proportion of these separated conversions implies a mechanistic association between the reciprocal and nonreciprocal exchanges. One mechanism would involve a reciprocal exchange to one side of the gene accompanied by a conversion "patch" encompassing both mutations, with heteroduplex present at both mutation sites. In envisioning such an intermediate, the specific initiating event is not implicit. Such an intermediate is consistent with either a single-strand invasion model (11), or with a double-strand, break-repair model, in which the gap is bordered by heteroduplex (12). With such a hypothetical intermediate, the observed products could be generated in either of two ways: by active correction of the distal mutation to wild type (conversion) accompanied by correction of the proximal mutation to wild

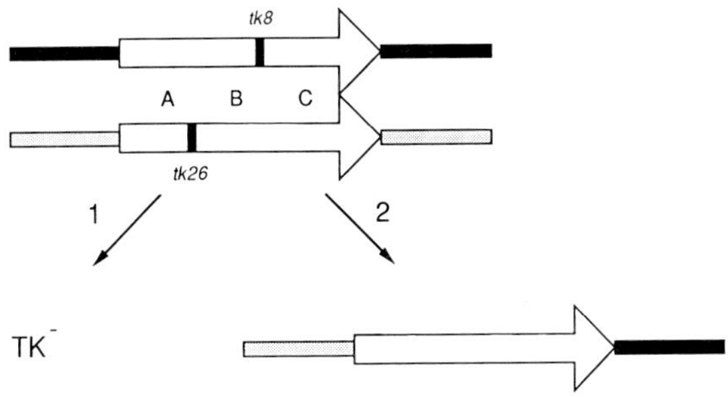

Fig. 4. Schematic representation of the product of reciprocal exchange accompanied by separated gene conversion. Flanking sequences are distinguished for the two alleles (filled for *tk8*, stippled for *tk26*). For orientation II direct repeats, simple reciprocal exchange in region B generates the products depicted in Fig. 3. Intrachromatid exchange of this type leads to pathway 1, and no products are recovered (TK$^-$). The wild-type product with flanking markers, as depicted for pathway 2, cannot be generated by simple reciprocal exchange. This product can be generated either by reciprocal exchange in A with conversion of *tk8* to wild type or by reciprocal exchange in C with conversion of *tk26*. In either case, the proximal mutation is not converted, as discussed in the text. For orientation II direct repeats, the product of pathway 2 would be a G418^s deletion product, which could be generated either through intrachromatid or through sister chromatid interactions. For orientation I direct repeats, the product of pathway 2 would be a triplication product for sister chromatid exchange; the product of intrachromatid exchange would loop out and would not be recovered.

type using the recipient strand as template (restoration), implying mixed correction; or by active correction of the distal mutation to wild type, but with no correction of the proximal mutation. The heteroduplex at the proximal site could then be resolved by the subsequent round of DNA replication, such that one sister chromatid is mutant, whereas the other is wild type and generates the observed deletion product. Because previous studies of chromosomal recombination from our laboratory failed to detect mixed correction products (13,14), we favor the latter scenario.

The deletion products in orientation II thus provide evidence that heteroduplex DNA can indeed be an intermediate in recombination, and that this heteroduplex can be actively corrected by the recombination machinery. A genetic analysis of G418 sensitivity suggests that greater than 3% of all orientation II products represent deletions. In the present study, a total of 11 independent G418^s recombinants were obtained, suggesting that these deletions do indeed represent a significant proportion of the products. Since reciprocal exchanges represent less than 20% of all products, deletion products must represent greater than 15% (3 of 20) of all reciprocal exchanges, making these a significant constituent class among reciprocal products. Further evidence for the prevalence of this type of process is provided by a single recombinant obtained between inverted repeats [one of 12 reciprocal exchanges (2)], and by triplication products of orientation I direct-repeat recombination; these products are under current analysis. Such separated gene conversions, associated with reciprocal exchange, orientation II deletions, and orientation I triplications, have been observed in analogous direct-repeat studies in yeast (6,9), and are commonly found during meiotic recombination in various fungi [see (15) for discussion]. Finally, these separated conversions have encouraged us to look more directly for the presence of heteroduplex DNA associated with intrachromosomal gene conversion in mammalian cells.

CONCLUSION

Studies with repeated *tk* genes in mouse L cells are providing insight into properties of mitotic recombination. In this investigation, comparisons of recombination products between alternate orientations of direct repeats have allowed an assessment of the relative contributions of sister chromatid and intrachromatid interactions. Although gene conversions represent the predominant product, these nonreciprocal exchanges are not informative regarding the type of interaction involved. By contrast, reciprocal exchanges resulting in changes in gene number provide a means to identify the interaction involved. A preponderance of triplications arising from sister chromatid exchange between orientation II direct repeats over presumed deletions arising from intrachromatid exchanges between orientation I direct repeats leads to the conclusion that sister chromatid interactions predominate in this system. The recovery of deletions between orientation II direct repeats, not expected to arise through simple reciprocal exchanges, provide evidence that reciprocal exchange can be accompanied by separated gene conversions. These products imply that recombination may involve a heteroduplex DNA intermediate. An association between these two modes of exchange, reciprocal and nonreciprocal, has been observed previously and suggests that the bias of sister chromatid strand interactions observed here among reciprocal exchanges may extend to gene conversions as well.

ACKNOWLEDGMENTS

This investigation was supported by Public Health Service Grant RO1 GM 32741 to R. M. L. R. J. B. was supported by a National Research Service Award (GM 07499) to Yale University. R. M. L. is a Leukemia Society of America Scholar.

REFERENCES

1. Liskay, R. M., Stachelek, J. L., and Letsou, A. (1984). *Cold Spring Harbor Symp. Quant. Biol.* **49,** 183–189.
2. Bollag, R. J., and Liskay, R. M. (1988). *Genetics* **119,** 161–169.
3. Letsou, A., and Liskay, R. M. (1987). *Genetics* **117,** 759–769.
4. Letsou, A., and Liskay, R. M. (1986). *In* "Gene Transfer" (R. Kucherlapati, ed.), pp. 383–409. Plenum, New York.
5. Maloney, D. H., and Fogel, S. (1987). *Curr. Genet.* **12,** 1–7.
6. Klein, H. L. (1988). *Genetics* **120,** 367–377.
7. Schiestl, R. H., Igarashi, S., and Hastings, P. J. (1988). *Genetics* **119,** 237–247.
8. Rothstein, R., Helms, C., and Rosenberg, N. (1987). *Mol. Cell. Biol.* **7,** 1198–1207.
9. Jackson, J. A., and Fink, G. R. (1981). *Nature (London)* **292,** 306–311.
10. Roman, H., and Fabre, F. (1983). *Proc. Natl. Acad. Sci. U.S.A.* **80,** 6912–6916.
11. Meselson, M. S., and Radding, C. M. (1975). *Proc. Natl. Acad. Sci. U.S.A.* **72,** 358–361.
12. Szostak, J. W., Orr-Weaver, T. L., Rothstein, R. J., and Stahl, F. W. (1983). *Cell* **33,** 25–35.
13. Liskay, R. M., and Stachelek, J. L. (1986). *Proc. Natl. Acad. Sci. U.S.A.* **83,** 1802–1806.
14. Waldman, A. S., and Liskay, R. M. (1988). *Mol. Cell. Biol.* **8,** 5350–5357.
15. Whitehouse, H. L. K. (1982). "Genetic Recombination: Understanding the Mechanisms." John Wiley, New York.

2

Studies on Extrachromosomal Homologous Recombination in Mammalian Cells: Implications for Chromosomal Recombination and Gene Targeting

MICHAEL LIN, KAREN SPERLE, AND NAT STERNBERG
Central Research and Development Department
Dupont-Merck Pharmaceuticals Company
Wilmington, Delaware 19880

INTRODUCTION

In recent years gene transfer techniques have been used extensively to study homologous recombination in cultured mammalian cells. Typically, one transfers one or two defective copies of a gene on the same or different DNAs into a cell, and then measures the conversion of those genes to wild type. The process can be analyzed genetically or physically either before the DNAs have stably integrated into the cellular genome, during that integration process, or after the DNAs have become stably integrated. Studies such as these indicate that the efficiency of recombination varies in a manner that seems to depend on the environment of the substrates that are used. Thus, homologous recombination between genes soon after they are transferred into cells, and before they have integrated into chromosomal DNA (extrachromosomal recombination), is very efficient. Frequencies of 1 to 10% that of wild-type transformation are routinely observed (1–6). This class of recombination events also appears to be promiscuous in that it is associated with a high frequency of mutation and with a level of illegitimate recombination that is about the same as the level of homologous recombination (7–8). A second class of recombination events that has been studied (chromosomal recombination) is between genes that have become stably integrated into chromosomes. These events are relatively inefficient (frequencies of 10^{-5} per cell generation are routine) and are almost always faithful in that they are rarely associated with either mutation or illegitimate rearrangement (9,10). Falling somewhere between these two classes of recombination is a third class of recombination events (gene targeting) that occurs between DNA introduced into cells and homologous chromosomal loci. Like class 2 recombination events, gene-targeting recombination is relatively inefficient, but the recombination frequency seems to vary considerably more than does recombination between two chromosomal genes. When the introduced DNA and the chromosome share about 10 kilobase pairs (kbp) of homology, the recombination frequencies range from 10^{-5} (targeted events per surviving cell) for the *hprt* and $C\mu$ loci to 3×10^{-7} and 2×10^{-9} for the *int-2* and *int-1* loci, respectively (11–13). Gene-targeting events are occasionally associated with the production of mutations (14), and are overwhelmed by the large excess of illegitimate integration events that introduce foreign DNA into nonhomologous chromosomal loci (3,11,15–16).

In addition to the environment of substrate DNA(s), the physical state of that DNA appears to affect its ability to recombine. Thus, Folger *et al.* (17) showed that within the first 60 min after DNA is injected into cells, it can recombine efficiently in a process that mimics extrachromosomal recombination. However, at later times, but still before it integrates into chromosomes, it takes on a form that resembles chromosomal DNA, and becomes refractory to both homologous and illegitimate recombination. It is noteworthy that recombination between the DNAs of lytically growing animal viruses resembles efficient extrachromosomal recombination. Whereas the nature of the change(s) that renders DNA refractory to recombination is not known, that process seems to be influenced, at least for extrachromosomal DNA, by the topological state of the introduced DNA (2; see the following).

In this chapter, we review some of our studies of extrachromosomal recombination in both mouse and human cells. The results indicate that linear

DNA is a much better substrate for homologous recombination than is circular DNA, and they suggest a nonconservative pathway for these events. We shall discuss whether this class of homologous recombination can serve as a model for chromosomal recombination and gene targeting.

RESULTS

Extrachromosomal Recombination in Mammalian Cells Is Stimulated by Cleaving the DNA Introduced into Those Cells at Appropriate Locations

The substrate that we use to illustrate this point is ptk^{2}DR (Fig. 1). Other substrates that give similar results are described in Lin *et al.* (18). ptk^{2}DR is a pBR322-based plasmid containing two defective herpes thymidine kinase genes that are oriented in the same direction. One of the genes contains a deletion at its 3′ end, and the other, a deletion at its 5′ end. They both share 320 bp of *tk* homology that are separated from each other on one side (the region proximal to homology or PH) by 1.5 kbp, and on the other side (the region distal to homology or DH) by 5.8 kbp. After this DNA is introduced into mouse *Ltk*$^-$ cells, Tk$^+$ transformed colonies are detected at a frequency that is no better than 0.02% that observed with a circular plasmid containing an intact *tk* gene (ptk) (Table I, lines 1 and 3). Linearizing the *tk* gene by cleaving it away from the vector with *Bam*HI has only a small negative effect on transformation (Table I, line 2). In contrast, linearizing the ptk^{2}DR DNA in the PH region of the plasmid at a *Hin*dIII site stimulates recombination more than 400-fold (Table I, line 4). This effect is seen only if the ends of the linear DNA are located in the PH region of the plasmid. Linearizing the plasmid at a *Sal*I site in the DH region of the plasmid has little effect on *tk* recombination (Table I, line 5). We have interpreted these and other similar results (2) to be supportive of a single-stranded annealing (SSA) model in which the ends of the linear molecule act as substrates for a 3′-exonuclease that degrades one strand of the DNA in either direction until complementary *tk* sequences are exposed (2; see also Fig. 4B). After these sequences pair, noncomplementary single-stranded regions are destroyed by nuclease activity, and a gap-repair process generates a wild-type gene. As predicted by this model, Southern analyses of Tk$^+$ transformants generated by *Hin*dIII-cleaved ptk^{2}DR DNA contain an intact *tk* gene, but are missing the PH region of the plasmid (18). Recent experiments of Maryon and Carroll (19) have shown that *Xenopus* oocyte nuclei, which can carry out recombination events

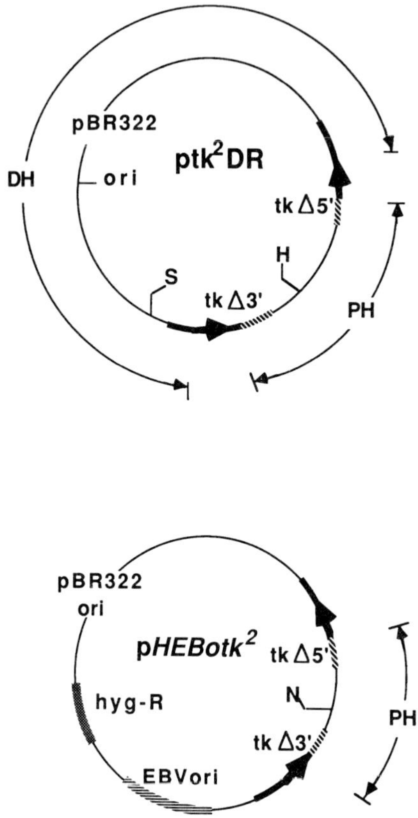

Fig. 1. DNA substrates used to assay intramolecular homologous recombination. Both of the plasmids contain two defective *tk* genes. The PH (proximal to homology) and DH (distal to homology) regions are also shown. The PH region is so designated because it is closer to the *tk* deletion junctions and to the shared *tk* homology than it is to the *tk* sequences not shared by the two defective *tk* genes. The DH region has just the opposite properties. Both plasmids contain the ampicillin-resistance gene, and the pBR322 plasmid replicon. p*HEBotk*^{2} also contains the hygromycin resistance gene and the EBV oriP. ———, pBR322 and nonherpes DNA; ▬, *tk* DNA; ➧, shared *tk* homology; ▨, deleted *tk* DNA. The arrowhead also indicates the direction of *tk* transcription. H, *Hin*dIII; N, *Not*I. ptk^{2}DR is described in greater detail in Lin *et al.* (18).

similar to those described here, appear to generate recombination intermediates that can be accounted for by exonuclease activity. In this case the exonuclease appears to be a 5′-exonuclease rather than a 3′-exonuclease. Wake *et al.* (20) proposed an alternative to the SSA model in which strands are not degraded at all but are rather unwound by a DNA helicase. This model can also account for the results reported here.

TABLE I
Effects of Double-Stranded Breaks on Intramolecular Recombination

DNA[a,b]	Restriction site(s) cleaved	DNA used (ng)	Tk^+ colonies/plates
p*tk*	none	10	2,890
	*Bam*HI	10	1,717
p*tk*²DR	none	50	0.7
	*Hind*III	50	409
	*Sal*I	50	3.7
p*HEBotk*	none	200	2,030
p*HEBotk*²	none	200	2
p*HEBotk*²	*Not*I	200	247

[a]The various p*tk*²DR or *ptk* DNAs were transformed into petri dishes containing 10^6 mouse *LtK*⁻ cells as calcium phosphate precipitates, and Tk^+ transformants were selected and scored in HAT medium. Each data point represents the average of three transformation plates. The structure of *ptk*²DR is shown in Fig. 1. *ptk* contains the herpesvirus 3.4-kbp *Bam*HI fragment cloned in pBR322 at the plasmid *Bam*HI site.

[b]The various p*HEBotk* or p*HEBotk*² DNAs were transformed in to 10^6 human 143E cells by electroporation, and Tk^+ transformants, selected in HAT medium. The plating efficiency after electroporation was 10–15%. Each data point is the average of two different transformations.

The Properties of Extrachromosomal Recombination Are Independent of the Cell Type Used, the Means of Transformation, and the Need for Stable Integration of the Transforming DNA

The experiments described above were carried out by calcium phosphate-mediated transformation of mouse L cells and require that the product of recombination become stably integrated in the cell's chromosome. To generalize the significance of those results, we measured recombination when DNA was introduced into cells by means other than calcium phosphate precipitation and when recombinant products were in their extrachromosomal state. To accomplish both of these goals, *tk* recombinants were scored after the transformation of human cell line 143–98.2 with p*HEBotk*² DNA (Fig. 1) (21–23). p*HEBotk*² DNA contains a hygromycin-resistance gene (*hyg*), an Epstein–Barr virus (EBV) plasmid origin of replication, and a tk² cassette that consists of an upstream *tk*Δ3′ gene with an SV40 promoter, a downstream *tk*Δ5′ gene, a 0.3-kbp intergenic segment of DNA between those genes, and a unique *Not*I site within the intergenic segment. Cell line 143–98.2 was generated from Tk⁻ cell line 143B by inserting a copy of the EBV EBNA-1 gene into the cellular genome. That gene produces a protein that permits the extrachromosomal replication of circular DNA containing the EBV plasmid ori (oriP). When p*HEBotk*^{2} DNA transforms 143–98.2 cells, the efficiency of stable transformation as measured by the production of *hyg*-R colonies (10^4 transformants generated/μg plasmid DNA) is 100–1000 times higher than it is with the parent 143B cell line that lacks the EBNA-1 gene. Presumably transformants generated in the latter case require stable integration of the introduced DNA, which is a much less efficient process than is the establishment of an extrachromosomal plasmid. This conclusion is supported by Southern analyses of DNA derived from those transformants. Transformants of 143–98.2 contain only free circles at a copy number 1–5 per cell, whereas transformants of 143B contain exclusively integrated p*HEBotk*^{2} DNA (data not shown). When 143–98.2 is transformed with circular p*HEBotk*^{2} DNA by electroporation the efficiency of Tk$^+$ recombination [HAT^r (hypoxanthine, aminopterin, thymidine-resistant) cells per surviving cell transformed] is very low (1.4×10^{-5}) (Table I). In contrast, if 143–98.2 is transformed with a p*HEBotk*^{2} DNA that has been cleaved in the PH region of the plasmid at the *Not*I site between the two defective *tk* genes, Tk recombination increase about 100-fold relative to uncleaved DNA. Southern analyses of DNA from transformants obtained with *Not*I-digested DNA indicate that this DNA is extrachromosomal, contains an intact *tk* gene, and is missing the intergenic segment between the defective *tk* genes in p*HEBotk*^{2} (Fig. 2). These results mimic the L cell transformation experiments and suggest that the observations made in both cases are independent of the cell line used (mouse or human), the method of transformation (calcium phosphate precipitation or electroporation), and the need to integrate the recombinant product. Recently, we have demonstrated elevated recombination levels with *Not*I-digested p*HEBotk*^{2} DNA that was introduced directly into 143–98.2 nuclei by microinjection (data not shown), indicating that the transforming DNA need not to go through the cytoplasm of the cell to behave in the manner described here. Seidman and co-workers (24,25) have obtained recombination results similar to those reported here by analyzing the extrachromosomal DNA population, following transformation of African green monkey kidney cells by the dextran sulfate method.

Extrachromosomal Recombination between DNAs Introduced into Mouse L Cells Is Mediated by a Nonconservative Pathway That Almost Exclusively Leads to Crossover Products

The experiments described in the previous section support the contention that extracellular recombination occurs by a degradative, nonconservative pathway that results in the loss of substrate sequences as the recombinant product is produced. To confirm this suggestion and to assess the relative

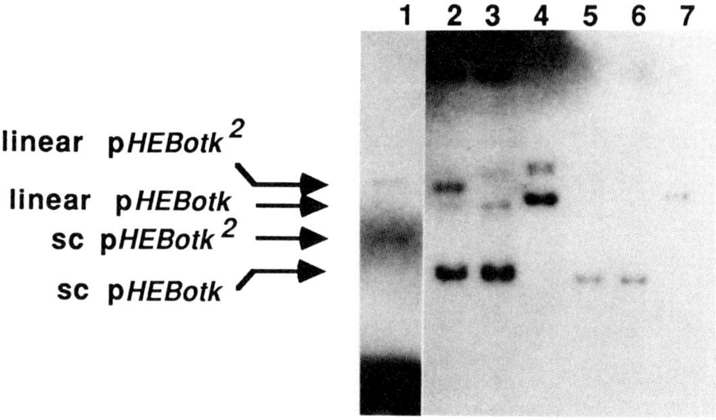

Fig. 2. Southern analysis of DNA derived from Tk$^+$ recombinants generated after p*HEBotk*^{2} transformation of 143–98.2. The Southern blotting protocol is described in Lin and Sternberg (1). Lane 1 contains both supercoiled (sc) and linear (*Xho*I–digested) p*HEBotk*^{2} DNA. Lanes 2–4 contain DNAs from one Tk$^+$ transformant, and lanes 5–7 contain DNAs from a second Tk$^+$ transformant. DNAs in lanes 2 and 5 are digested with *Xba*I; DNAs in lanes 3 and 6 are digested with *Bgl*II; DNAs in lanes 4 and 7 are digested with *Xho*I. *Xba*I and *Bgl*II do not cleave p*HEBotk*^{2} or the p*HEBotk* recombinant DNA. *Xho*I cleaves these DNAs once. Nick-translated p*HEBotk*^{2} DNA was used as probe.

contribution of the degradative SSA model and the double-strand break repair (DSBR) model (26) to the extrachromosomal recombination process in cultured mammalian cells, a series of intermolecular recombination experiments was performed. The *tk* plasmids used in these experiments are shown in Fig. 3. p*tk*B8 is a plasmid containing an inactive *tk* gene with an 8-bp *Bam*HI linker insertion at a central *Sph* I site within *tk*. Mp10*tk*Δ3′, mp10*tk*Δ5′, and mp10*tk*Δ3′Δ5′ are double-stranded, supercoiled mp10 substrates containing *tk* genes deleted either at their 3′ or 5′ ends or at both ends. *Ltk*$^-$ cells were transformed with p*tk*B8 DNA, and one of the mp10 DNAs by the calcium phosphate precipitation procedure; Tk$^+$ recombinants were scored after transfer of the cells to HAT medium. The results of some of these experiments are shown in Table II, and the results of others are summarized in Lin *et al.* (2). Several conclusions can be drawn from these studies: (1) If circular p*tk*B8 DNA is used with either circular or linear versions of any of the mp10*tk* DNAs (Table II, lines 1,2,7), then the recombination frequency is very low (fewer than 50 transformants per 10^6 cells transformed are detected in any case); (2) If p*tk*B8 is cleaved at the *Bam*HI site in tk and that DNA is used to transform cells along with circular mp10*tk*Δ3′Δ5′ DNA (Table II, line 3), no transformants are detected. Assuming recombination intermediates generated in the DSBR model can be

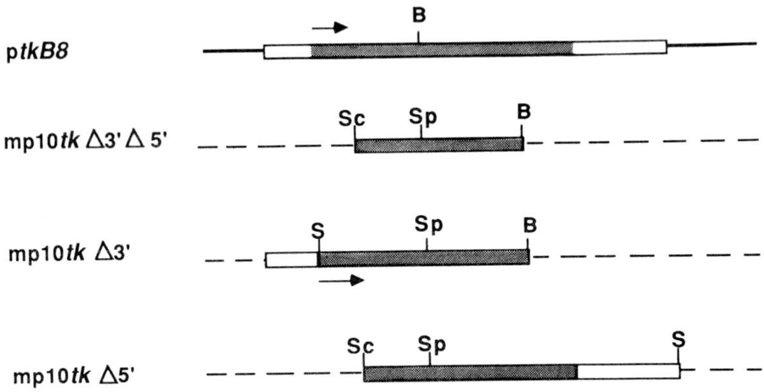

Fig. 3. Mutant *tk* genes used for studies of intermolecular homologous recombination. ▬ , Herpesvirus *tk* DNA; ———, pBR322 plasmid DNA; – – – –, mp10 DNA; ➧, direction of *tk* transcription. B, *Bam*HI; Sp, *Sph*I; Sc, *Sac*I; S, *Sal*I. Further detailed maps of these genes and their plasmids are found in Lin *et al.* (2).

resolved equally well to crossover or noncrossover products, this model predicts that the latter combination of substrates should easily generate Tk^+ recombinants—the mp10*tk*Δ3′Δ5′ DNA should act as a donor of information to reconstruct the defective linear *tk*B8 DNA (2,26). Since the SSA model predicts that one should produce only crossover recombination products, and mp10*tk*Δ3′Δ5′ is defective at both ends of the gene, Tk^+ recombinants should not be generated with these substrates by this recombination pathway; (3) If *Bam*HI-cleaved p*tk*B8 DNA is used with Mp10*tk*Δ3′Δ5′ DNA (Table II, line 4 and 5) that has been linearized at either end of the doubly defective gene, Tk^+ recombinants are produced at a frequency only slightly higher than that seen with circular mp10*tk*Δ3′Δ5′ DNA (10–30 transformants per 10^6 cells). In contrast, if the doubly defective *tk* gene is cleaved away from the mp10 vector DNA, the recombination frequency is increased 10-fold relative to the singly cut mp10*tk*Δ3′Δ5′ DNA (Table II, line 6). We have interpreted the latter result according to the SSA model shown in Fig. 4A (27). In this model, 3′ exonuclease degradation from the *Bam*HI ends of p*tk*B8 DNA and from the ends of the double-defective *tk*Δ3′Δ5′ fragment generates complementary single-stranded regions, which can pair to insert wild-type *tk* sequences from tkΔ3′Δ5′ DNA into p*tk*B8 DNA in place of the *tk*B8 mutation. Subsequent gap repair and ligation reactions reconstructs a wild-type *tk* gene; (4) If *Bam*HI-cleaved p*tk*B8 DNA is transformed into cells along with either mp10*tk*Δ3′ or mp*tk*Δ5′ DNA that has been cleaved at the junction between the deletion mutation and the vector DNA (*Bam*HI-cleaved Δ3′ DNA or *Sac*I-cleaved Δ5′, Table II, lines 9,10,14) then the

TABLE II
Effect of Double-Stranded Breaks on Intermolecular Recombination between p*tk*B8, *tk*Δ3′Δ5′, *tk*Δ3′, and *tk*Δ5′ DNAs[a]

	p*tk*B8 DNA		mp10*tk*Δ3′Δ5′			
Line	Amount (ng)	Enzyme treatment	DNA	Amount (ng)	Enzyme treatment	Tk⁺ colonies per plate
1	100	none	mp10*tk*Δ3′Δ5′	100	none	1
2	100	none		100	*Sac*I + *Bam*HI	1
3	100	*Bam*HI		100	none	0
4	100	*Bam*HI		100	*Sac*I	33
5	100	*Bam*HI		100	*Bam*HI	7
6	100	*Bam*HI		100	*Sac*I + *Bam*HI	312
7	100	none	mp10*tk*Δ3′	100	none	47
8	10	*Bam*HI		10	none	6
9	10	*Bam*HI		10	*Bam*HI	340
10	100	*Bam*HI		100	*Bam*HI	2410
11	10	*Bam*HI		10	*Sal*I	2
12	100	*Bam*HI		100	*Sal*I	191
13	100	*Bam*HI	mp10*tk*Δ5′	100	none	18
14	100	*Bam*HI		100	*Sac*I	1298

[a]Transformation was carried out by the calcium phosphate precipitation method as in Table I, footnote *a*.

recombination level increases dramatically to more than 1000 transformants per 10^6 cells. In contrast, if mp10*tk*Δ3′ DNA is cleaved at its other *tk*-vector junction (e.g., at a *Sal*I site) then recombination is 10- to 100-fold less efficient (Table II, lines 9–12). The SSA model that can account for the results with *Bam*HI-digested p*tk*B8 and *Bam*HI-digested mp10*tk*Δ3′ DNAs is shown in Fig. 4B. Again, 3′-exonuclease digestion from the ends of both of these DNAs exposes complementary *tk* sequences that can pair and be repaired to reconstruct a wild-type *tk* gene. A clear prediction of this model is that all the recombinants should be of the crossover type. Southern analyses of DNAs derived from transformants generated with *Bam*HI-digested p*tk*B8 and *Bam*HI-digested mp10*tk*Δ3′ DNAs indicate that 29 of 30 Tk⁺ transformants contain *tk* genes generated by a crossover recombination event (2). Moreover, very few of the transformants contain the original *tk* transforming DNAs, indicating that those DNAs have likely been destroyed in the recombination process. The nonconservative properties of extrachromosomal recombination are supported also by experiments from several other laboratories (6,24–25).

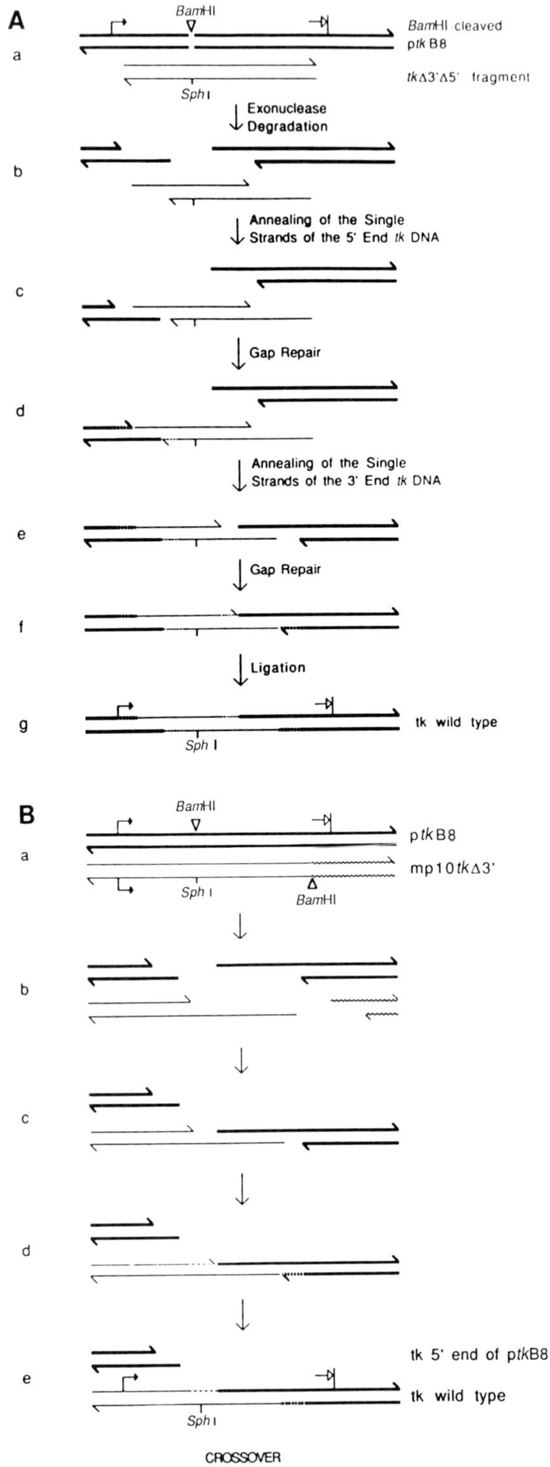
A
a
BamHI
BamHI cleaved
ptk B8
tkΔ3'Δ5' fragment
Sph I
Exonuclease
Degradation
b
Annealing of the Single
Strands of the 5' End tk DNA
c
Gap Repair
d
Annealing of the Single
Strands of the 3' End tk DNA
e
Gap Repair
f
Ligation
g
tk wild type
Sph I
B
a
BamHI
ptkB8
mp10tkΔ3'
Sph I
BamHI
b
c
d
e
tk 5' end of ptkB8
tk wild type
Sph I
CROSSOVER

DISCUSSION

One clear observation that emerges from our experiments on extrachromosomal recombination is the fact that circles are extremely poor substrates for recombination, even when only one of the two substrates in an intermolecular reaction is circular. We can think of two possible reasons for this. First, DNA ends are needed to initiate the recombination process and must be present on both of the substrates that undergo intermolecular reactions. This possibility is more easily accommodated by the SSA model than the DSBR model. Another reason for the nonrecombinogenic nature of circular DNA may be due to the fact that it becomes refractory to recombination more rapidly than does linear DNA after either DNA is introduced into mammalian cells. If one hypothesizes that this refractory state is associated with, for example, the conversion of naked DNA to chromatin, these results would imply that circular DNA is converted to chromatin more rapidly than linear DNA following transformation. Moreover, if circular DNA is such a poor recombination substrate, then the efficiency with which linear DNA is ligated to form a circle after it enters a cell is likely to influence significantly the efficiency of recombination. Perhaps these observations can explain the variability with which introduced circular and linear DNAs target homologous chromosomal loci. In some experiments, linear DNAs were shown to be more efficient than circles (11,14,28), and in other experiments, the two DNAs were shown to target chromosomal genes with the same efficiency (29). One wonders whether differences in the ligation capacities of the cells used in the two experiments could account for the contrasting results.

Is the SSA model a relevant model for chromosomal and targeting recombination reactions? As many, if not most, of the recombination products generated by these processes are non-crossover products that appear to result from conservative, non-degradative processes, the importance of SSA pathway in these reactions remains a real question. Since this pathway appears not to be simply an artifact of the DNA transformation process (it is independent of the cell line used, the transformation method used, and the

---

Fig. 4. The single-strand annealing model. In both of the models, strand degradation is shown to be mediated by a 3′-endonuclease. Similar models mediated by a 5′-exonuclease are also possible (1). (A) This pathway describes the proposed steps in the efficient correction of *Bam*HI-cleaved p*tk*B8 DNA with an internal *tk*Δ3′Δ5′ fragment. Thick and thin lines represent *tk* DNA from p*tk*B8 and internal *tk* fragment, respectively. pBR322 sequences in p*tk*B8 are not shown. Broken lines indicated newly synthesized DNA generated during gap repair. (B) This pathway shows how *Bam*HI-cleaved p*tk*B8 and mp10Δ3′ DNAs can generate an intact *tk* gene in which flanking markers are rearranged. Symbols are as in part A. Wavy lines represent nonhomologous mp10 DNA. pBR322 DNA in p*tk*B8 is not shown.

need for DNA to integrate into the genome of the cell), it is worth considering more carefully its possible role in chromosomal recombination. Indeed, one could argue that the degradative nature of the SSA pathway frequently prevents its detection in chromosomal recombination events. For example, for intrachromosomal recombination, the double-stranded break in chromosomal DNA that is necessary to produce a fruitful reaction would have to be appropriately located relative to the homologous regions undergoing recombination. The likelihood that this might happen could be low. Indeed, according to this scenario, one might suspect that the chromosomal locus at which this pathway might operate best is one containing tandemly repeated DNA. This suggestion is noteworthy, in that Ozenberger and Boeder (30) have shown that recombination products detected at the tandemly repeated ribosomal DNA (rDNA) locus in *Saccharomyces cerevisiae* are more readily accounted for by an SSA model than a DSBR model. Moreover, Ray *et al.* (31) have described a pathway for recombination initiated by a double-stranded cut near the *his3* locus in *S. cerevisiae*, in which long regions of single-stranded DNA are presumably generated from the DNA ends. It is argued that these regions take part in homologous pairing and generate recombinants, but that the pathway is not a degradative one in that the chromosome is restored by resynthesis of the degraded strand. Such a modified SSA pathway might explain the conservative recombination products seen in most mammalian chromosomal recombination events. As far as interchromosomal recombination is concerned, it seems obvious why a strict SSA model might be inefficient, since double-stranded cuts would have to be made in both chromosomes that take part in the recombination. This contrasts with the DSBR model in which a double-stranded cut need be made in only one chromosome.

What are the implications of these results for gene targeting? One clear implication is that the efficiency of targeting might be significantly improved using a model similar to that shown in Fig. 4A, if the introduced DNA were a linear fragment whose ends were homologous to the target locus, and if a double-stranded break could be produced in the homologous chromosomal locus. Since it has not been possible to cut chromosomes at specific loci, this possibility has not yet been tested. Perhaps the introduction of the MAT locus and HO endonuclease (32,33) into mammalian cells will permit the testing of this hypothesis. In this regard, it should be noted that targeting reactions are standardly carried out with introduced linear fragments that contain nonhomologous sequences at one or both ends (11). Indeed, it has been shown that such a substrate targets homologous chromosomal loci no less efficiently than does a fragment that is homologous to the chromosome at both of its ends (11). We can think of two explanations for these results: (1) Ends act not as sites of strand degradation but rather as entry points for a recombinase that moves along the DNA and generates recombination intermediates. If these intermediates are produced in a region of the fragment that shares homology with a chromosomal locus, a targeting event can occur. This scenario is analogous to that proposed for *rec*BCD-mediated recombination in *Escherichia coli* (34,35) and for certain double-strand, break-mediated recombination events in *S. cerevisiae* (36). (2) Targeting events occur in those cells in which the introduced linear DNA undergoes end degradation, such as that proposed in the SSA model in Fig. 4. That degradation would expose single-stranded regions homologous to chromosomal loci that are internal to the fragment, regardless of whether the ends of the fragment bear chromosomal homology. Both explanations are consistent with the low efficiency with which circles take part in these reactions.

It is clear from these discussions that we have much to learn about the mechanism of homologous recombination in mammalian cells. It is hoped that as we learn more, we can put that information to better use in generating specific gene mutations.

REFERENCES

1. Lin, F.-L., Sperle, K., and Sternberg, N. (1984). *Mol. Cell. Biol.* **4,** 1020–1034.
2. Lin, F.-L., Sperle, K., and Sternberg, N. (1990). *Mol. Cell. Biol.* **10,** 103–112.
3. Folger, K. R., Wong, E. A., Wahl, G., and Capecchi, M. R. (1982). *Mol. Cell. Biol.* **2,** 1372–1387.
4. Kutcherlapati, R. S., Eves, E. M., Song, K.-Y., Morse, B. S., and Smithies, O. (1984). *Proc. Natl. Acad. Sci. U.S.A.* **81,** 3153–3157.
5. Wake, C. T., and Wilson, J. H. (1979). *Proc. Natl. Acad. Sci. U.S.A.* **76,** 2676–2880.
6. Anderson, R. A., and Eliason, S. L. (1988). *Mol. Cell. Biol.* **6,** 3246–3252.
7. Roth, D., and Wilson, J. H. (1985). *Proc. Natl. Acad. Sci. U.S.A.* **82,** 3355–3359.
8. Anderson, R. A., Kato, S., and Camerini-Otero, R. D. (1984). *Proc. Natl. Acad. Sci. U.S.A.* **81,** 206–210.
9. Stachelek, J. L., and Liskay, R. M. (1988). *Nucleic Acids Res.* **16,** 4069–4076.
10. Waldman, A. S., and Liskay, R. M. (1987). *Proc. Natl. Acad. Sci. U.S.A.* **84,** 5340–5344.
11. Mansour, S. L., Thomas, K. R., and Capecchi, M. R. (1988). *Nature (London)* **24,** 348–352.
12. Shulman, M. J., Nissen, L., and Collins, C. (1990). *Mol. Cell. Biol.* **10,** 4466–4472.
13. Thomas, K. R., and Capecchi, M. R. (1990). *Nature (London)* **346,** 847–850.
14. Thomas, K. R., Folger, K. R., and Capecchi, M. R. (1986). *Cell* **44,** 419–428.
15. Roth, D. B., Porter, T. N., and Wilson, J. H. (1985). *Mol. Cell. Biol.* **5,** 2599–2607.
16. Adair, G., Nairn, R. S., Wilson, J. H., Seidman, M., Brotherman, K. A., Mackinnon, C., and Scheerer, J. B. (1989). *Proc. Natl. Acad. Sci. U.S.A.* **86,** 4574–4578.
17. Folger, K. R., Thomas, K., and Capecchi, M. R. (1985). *Mol. Cell. Biol.* **5,** 59–69.
18. Lin, F.-L., Sperle, K., and Sternberg, N. (1987). *Mol. Cell. Biol.* **7,** 129–140.
19. Maryon, E., and Carroll, D. (1989). *Mol. Cell. Biol.* **9,** 4862–4871.
20. Wake, C. T., Vernaleone, F., and Wilson, J. H. (1985). *Mol. Cell. Biol.* **5,** 2080–2089.
21. Sudgen, B., Marsh, K., and Yates, J. (1985). *Mol. Cell. Biol.* **5,** 410–413.

22. Yates, J., Warren, N., Reisman, D., and Sugden, B. (1984). *Proc. Natl. Acad. Sci. U.S.A.* **81,** 3806–3810.
23. Yates, J., Warren, N., and Sugden, B. (1985). *Nature (London)* **313,** 812–815.
24. Chakrabarti, S., and Seidman, N. M. (1986). *Mol. Cell. Biol.* **6,** 2520–2526.
25. Seidman, N. M. (1987). *Mol. Cell. Biol.* **7,** 3561–3565.
26. Szostak, J. W., Orr-Weaver, T. C., Rothstein, R. J., and Stahl, F. W. (1983). *Cell* **33,** 25–35.
27. Lin, F.-L., Sperle, K., and Sternberg, N. (1990). *Mol. Cell. Biol.* **10,** 113–119.
28. Jason, M., and Berg, P. (1988). *Genes Dev.* **2,** 1353–1363.
29. Adair, G., and Wilson, J. (1991). This volume
30. Ozenberger, B. A., and Roeder, G. S. (1991). *Mol. Cell. Biol.* **11,** 1222–1231.
31. Ray, A., Machin, N., and Stahl, F. W. (1989). *Proc. Natl. Acad. Sci. U.S.A.* **86,** 6225–6229.
32. Klar, A. J. S., Strathern, J. N., and Abraham, J. A. (1984). *Cold Spring Harbor Symp. Quant. Biol.* **49,** 77–88.
33. Kastriken, R., and Heffron, F. (1984). *Cold Spring Harbor Symp. Quant. Biol.* **49,** 89–104.
34. Kobayaski, I., Stahl, M. M., and Stahl, F. W. (1984). *Cold Spring Harbor Symp. Quant. Biol.* **49,** 497–506.
35. Smith, G. A., Amundsen, S. K., Chaudhury, A. M., Cheng, K. C., Ponticelli, A. S., Roberts, C. M., Schultz, D. W., and Taylor, A. F. (1984). *Cold Spring Harbor Symp. Quant. Biol.* **49,** 485–495.
36. Nickoloff, J. A., Singer, J. D., Hoekstra, M. F., and Heffron, F. (1989). *J. Mol. Biol.* **207,** 527–541.

3

Homologous Recombination in Embryonic Stem Cells as a Means to Generate Mice with Defined Mutations

ANDREAS ZIMMER[1,2], ZHAO-QI WANG[3],
ERWIN F. WAGNER[3], AND PETER GRUSS[1]
1. Max-Planck Institute for Biophysical Chemistry
Am Fassberg, 3400 Göttingen, Germany
and
2. Developmental Biology Unit, Laboratory of Cell Biology
National Institute of Mental Health
Bethesda Maryland, 20982
and
3. Research Institute of Molecular Pathology
Dr Bohr-Gasse 7, A- 1030 Wien, Austria

INTRODUCTION

The integration of DNA by homologous recombination in mammalian cells has many potential genetic and therapeutic applications. These include mutation of genes in embryonic stem (ES) cells for gene function analysis and gene therapy to repair defective genes in human somatic cells. Whereas the techniques are not yet advanced enough to be considered for gene therapy, the routine generation of mouse mutants will soon be the state of the art.

Embryonic stem cells are pluripotent cells derived from preimplantation mouse embryos (Martin, 1981; Evans and Kaufmann, 1981; Doetschmann *et al.*, 1985). They can be propagated and manipulated in culture and subsequently reintroduced into host blastocysts to form chimeric mice.

Eventually they contribute to the germ cell lineage in the chimeras (Bradley *et al.*, 1984; Robertson *et al.*, 1986; Gossler *et al.*, 1986). Thus, it is possible to introduce mutations via ES cells into the mouse germ line. These mutants should be very useful for analysis of gene function and to establish animal models for human genetic diseases.

One major obstacle in gene targeting in mammalian cells has been that illegitimate recombination is usually two to three orders more efficient than homologous recombination (reviewed in Capecchi, 1989). In spite of this problem, strategies have been developed to enrich for cells that bear homologously recombined alleles (Mansour *et al.*, 1988; Jasin and Berg, 1988). Unfortunately, only a few genes allow the direct screening for a loss-of-function phenotype, as for example the *hprt* gene (Hooper *et al.*, 1987). In most cases the target gene is interrupted with a marker gene, usually a *neo*^{r} gene. However, there are many cases in which the introduction of a marker is not desired. One might, for example, wish to introduce very specific mutations into protein coding or regulatory regions. Or there might be concern about the effect of the introduction of a complete gene with regulatory sequences into the target locus.

Here we summarize experiments in which we employed a new method to mutate the mouse homeobox *Hox-1.1* gene in ES cells (Zimmer and Gruss, 1989). This method does not require selection and should allow the introduction of small mutations without further alteration of the target locus.

GENERAL STRATEGY

All mammalian *Hox* genes are organized in clusters. These clusters are very similar to the *Drosophila antennapedia* and *Bithorax* homeobox clusters in their gene organization and pattern of expression (Gaunt *et al.*, 1988; Duboule and Dolle, 1989; Graham *et al.*, 1989). We assumed that this clustered organization might be important for gene regulation and that *cis* regulatory elements control the expression of several homeobox genes. It has been shown that the overexpression or ectopic expression of homeobox genes interferes with the normal embryonic development (Wohlgemuth *et al.*, 1989; Balling *et al.*, 1989; Kessel *et al.*, 1990). For this reason, we were concerned that disruption of the *Hox-1.1* gene with a *neo*^{r} gene, which would introduce a strong promoter into the *Hox-1* cluster, would lead to the disregulation of other genes in the cluster. This might make a proper analysis of the *Hox-1.1*$^-$ phenotype impossible. Thus, we decided to disrupt the *Hox-1.1* coding region by integrating a 20-basepair (bp) oligonucleotide into the homeobox.

In these experiments we could not select for cells carrying a mutated

Hox-1.1 allele, nor could we select for cells that have integrated the foreign DNA by illegitimate recombination. Crucial for the success of these experiments therefore was very efficient DNA transfection as well as highly sensitive and specific analysis. Finally, we had to develop a protocol that allowed us to clone mutated cells out of a pool of nonmutated cells.

The most efficient method to introduce DNA into eukaryotic cells is by microinjection into the nucleus (Capecchi, 1980). This was therefore the method of choice, and we used microinjection to transform pools of 100 to 250 cells.

The polymerase chain reaction (PCR) was then used to identify those cell pools that contain cells carrying the *Hox-1.1*$^-$ allele (Fig. 1). One primer used in the PCR was identical to the integrated oligonucleotide and was not present in the endogenous *Hox-1.1* gene. The other primer hybridized to *Hox-1.1* sequences outside the incoming DNA. Only after homologous recombination would the primers be linked and give rise to an indicative fragment in the PCR. As shown in Fig. 1, homologous recombination with the shortest construct pH/O-F could be detected by the amplification of a 5′

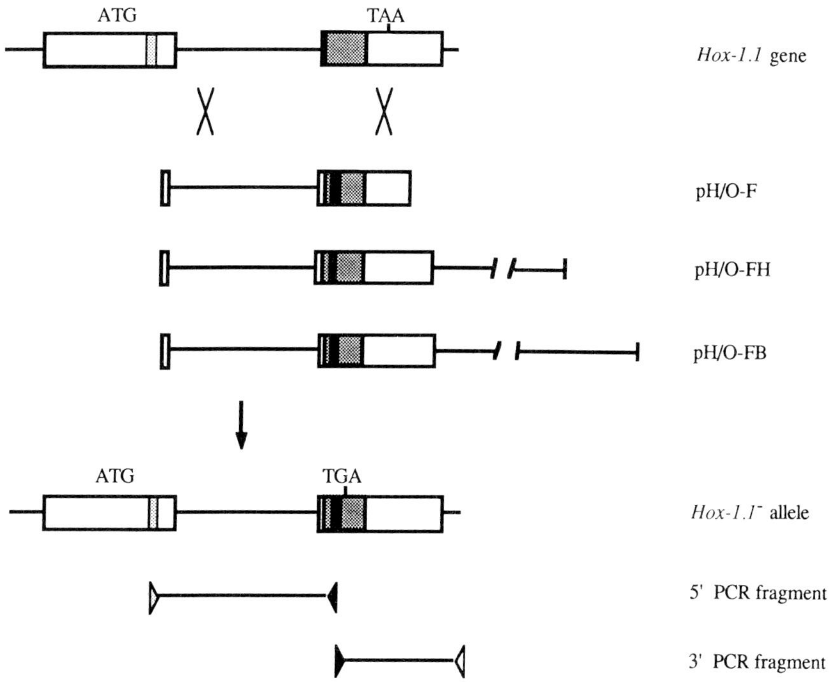

Fig. 1. Planned disruption of the *Hox-1.1* gene with the different targeting vectors and its detection by PCR. See text for details.

and a 3′ fragment. Homologous recombination events with the longer constructs were detectable only by amplification of the 5′ fragment.

Homologously recombined cells were cloned after the identification of positive pools by sib selection. First ES cell colonies were separated and analyzed by PCR. In the next step, individual ES cell were isolated from positive subclones to yield monoclonal cell lines. Because the oligonucleotide introduced a *KpnI* site at the expense of an *EcoRI* site, cell clones carrying the *Hox-1.1*$^{-}$ allele could be unambiguously identified by genomic Southern analysis.

MICROINJECTION OF MUTATED *HOX-1.1* FRAGMENTS INTO EMBRYONIC STEM CELLS

We used in all experiments the ES cell line D3 (Doetschmann *et al.*, 1985). D3 cells grow in densely packed colonies. Individual cells are usually difficult to distinguish. This makes the microinjection of D3 cells more difficult than the injection of most other cell types. In addition, it is necessary to inject as many cells as possible to keep the number and size of the starting pools small.

To test whether it would be feasible to inject a small population of D3 cells in a quantitative manner, we performed some pilot experiments in which we injected fluorescein-labeled bovine serum albumin-fluorescein isothiocyamate (BSA-FITC) into D3 cells under various conditions.

We found that it was best to inject the cells a few hours after plating, when the colonies are not yet fully packed. For this purpose, logarithmically growing populations were trypsinized and diluted into medium conditioned by buffalo rat liver (BRL) cells containing 5% fetal calf serum (FCS) to 2.5 to 5 × 10^4 cells/ml. Twenty-μl drops were placed in the middle of 60-mm tissue culture dishes in an area that had been previously gelatinized. The drops were than surrounded with regular ES medium containing 15% FCS in order to generate a humid chamber. The plates were then incubated at 37°C in 5% CO_2 for 3 to 8 hr. During this time, the cells attached to the dish and became suitable for injection. Only plates that contained between 50 and 200 cells growing in small colonies of 3 to 10 cells were used. Immediately before injection, additional ES medium was added, to a total of 8 ml. The cells were injected in the nucleus with minimal injection pressure. After the injections, 1×10^5 primary mouse fibroblasts were added to each plate as feeder cells.

Figure 2 shows the injection of BSA-FITC into a colony of D3 cells under optimum conditions. Five hr after the injection, the cells were fixed and stained with 4′, 6′-diamino-2-phenylindole (DAPI) to visualize the nuclei. All cells that have been injected exhibit a bright green fluorescence, whereas

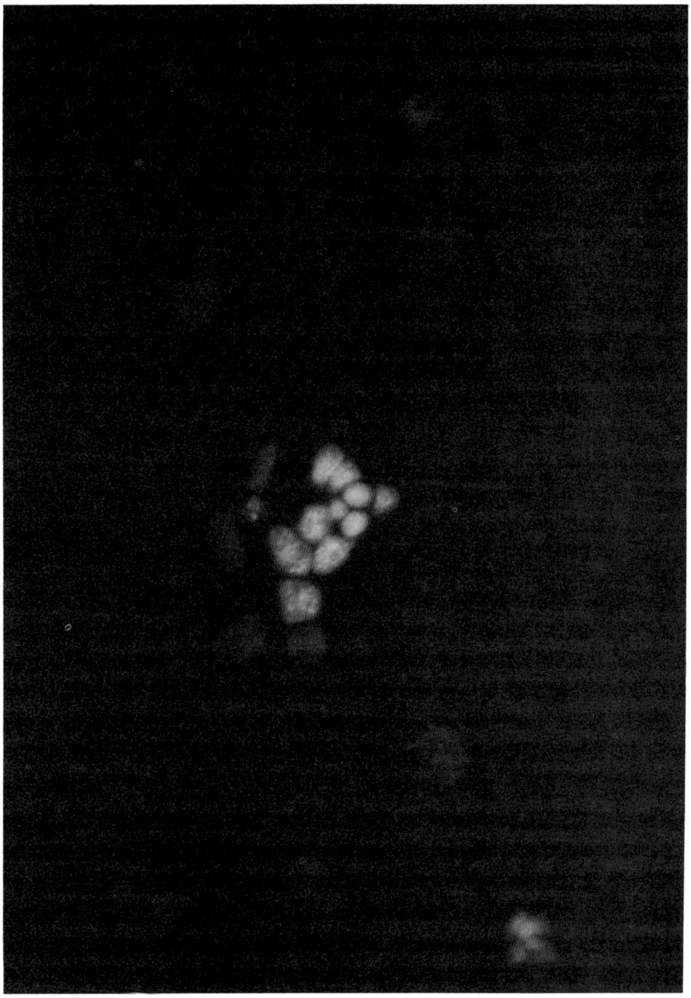

Fig. 2. Microinjection of BSA-FITC into an ES cell colony. All ES cells in a single colony were injected with BSA-FITC at a concentration of 2 mg/ml. Cells were fixed after 5 hr and stained with DAPI to visualize cell nuclei. Microinjected nuclei of ES cell exhibit a bright FITC-fluorescence, in contrast to the faint DAPI-fluorescence of noninjected surrounding fibroblast nuclei.

the noninjected feeder cells show only the blue DAPI staining. This demonstrates that with some experience, it is possible to inject all cells in an D3 cell colony.

Having established these conditions, we injected the excised fragments of three different constructs into D3 cells (see Fig. 1). The constructs differ in

TABLE I
Microinjection Experiment with Different Targeting Vectors[a]

Construct	No cell injected	PCR positive	Frequency
pH/O-F	739	5	1:150
pH/O-FH	400	3	1:130
pH/O-FB	669	6	1:110

[a]See Fig. 1. Generally not more than 50 cells per pool were injected.

the length of homology downstream of the homeobox: pH/O-F had a homology of 1.56 kb, pH/O-FH had a homology of 4 kb, and pH/O-FB a homology of 8.5 kb. DNA was isolated from injected pools, after the cells were grown for 2 weeks, and subjected to analysis. As outlined in Table I, we obtained the highest frequency of homologous recombination with those constructs that had the longest homology to the target locus. In total the frequency was surprisingly high, with about 1 in 110 injected cells with the longest and 1 in 150 cells with the shortest construct having undergone homologous recombination.

CLONING OF HOMOLOGOUS RECOMBINED CELL LINES

The next step was the isolation of cell lines with a homologously recombined *Hox-1.1* allele. First the cells were plated at low density on feeder cells, so that colonies formed after 3 to 4 days in culture. Under these conditions, the colonies were not of clonal origin, but had a much lower complexity than the original cell pool. These colonies were individually harvested and expanded. DNA isolated from the subcultures was first pooled and analyzed by PCR analysis. Those DNA pools that gave the strongest signal were then individually analyzed by PCR. Figure 3 shows an example of such an experiment. After the first amplification of the DNA pools, a

Fig. 3. Analysis of an experiment to subclone mutated ES cells after injection of the construct pH/O-F (A) genomic DNA from 5 subclones (100 ng each) was pooled and amplified by PCR for 30 cycles to yield the indicative 3′ fragment. The products were separated on a 1% agarose gel, transferred onto Gene Screen Plus and hybridized with a *Hox-1.1* P^{32}-labeled homeobox probe. Shown is the ethidium bromide-stained agarose gel and the autoradiogram. Notice the strong hybridization signal from the subclones 21–25. (B) Individual analysis of the subclones 21–25. DNA (100 ng each) was amplified for 35 cycles.

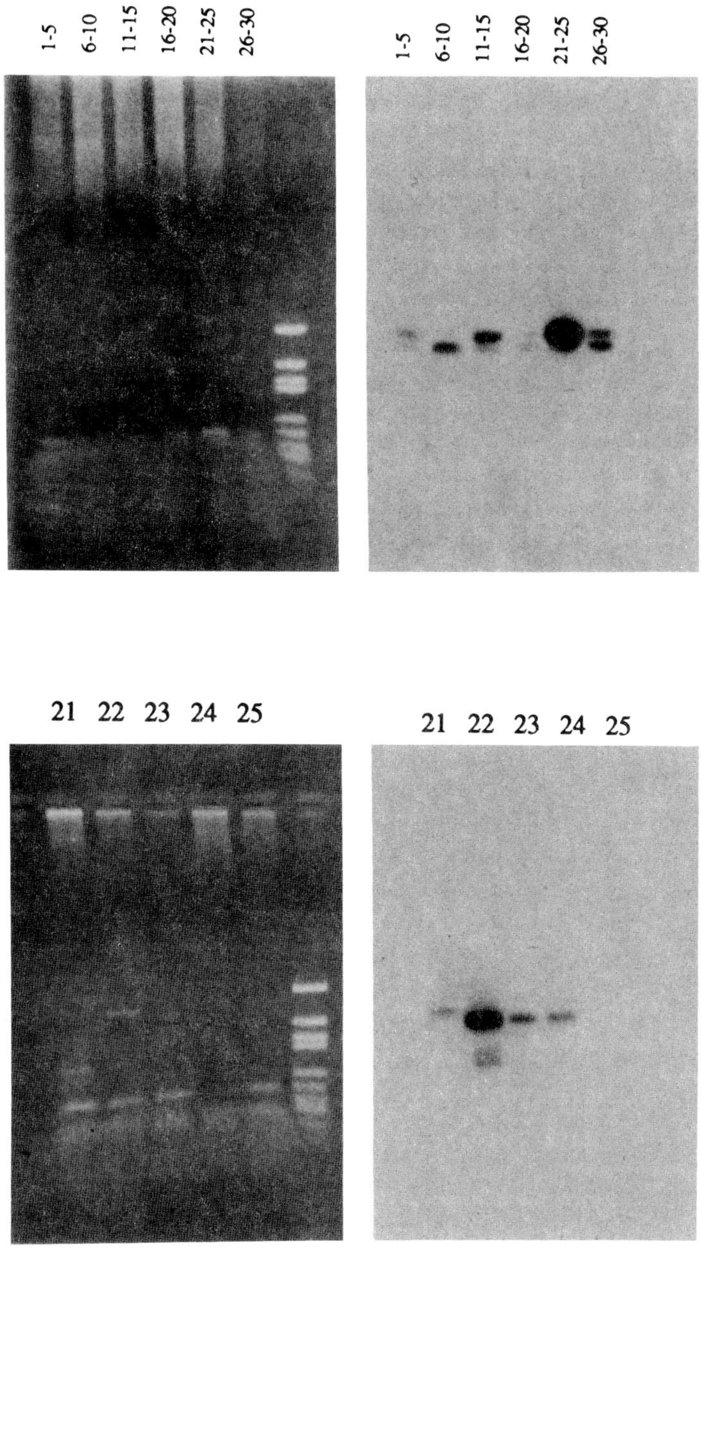
1-5
6-10
11-15
16-20
21-25
26-30
1-5
6-10
11-15
16-20
21-25
26-30
21 22 23 24 25
21 22 23 24 25

strong hybridization signal was detected, but the amplified fragment was not visible in the ethidium bromide-stained gel. However, if DNA from individual pools was amplified, the band indicative of a homologous recombination event was readily visualized in the ethidium bromide-stained gel.

In the next step we isolated individual cells and expanded them to form cell clones. These clones were then analyzed by genomic Southern blotting. Two independent mutated D3 cell lines (clones A90 and P4) were isolated by this means and selected for further characterization.

GENERATION OF CHIMERIC ANIMALS WITH HOMOLOGOUSLY RECOMBINED CELLS

The *Hox-1.1*$^{-}$ clones were injected into NMRI, MF1 and C57BI/6J blastocysts. Between 10 and 15 cells were injected into each blastocyst, and 10–30 blastocysts were retransferred into each pseudopregnant foster mother. The D3 cell line was originally derived from a 129/SV mouse strain (Doetschmann *et al.*, 1985). This strain is homozygous for the black (b) and the agouti (a) locus and exhibits the typical black-agouti coat color. Therefore, chimeric animals could be easily identified on the albino background of the MF1 and NMRI and on the nonagouti background of C57BI/6J hosts (Fig. 4).

The results are summarized in Table II. We injected a total of 681 MF1 and C57BI/6J blastocysts with the clone P4 and obtained 6 male chimeric animals. The overall chimerism in these males was poor and did not exceed 10%, as judged by coat color. The animals were mated with albino or C57BI/6J females. Five males were fertile, but from 138 offspring analyzed, none exhibited the black-agouti coat color.

We next injected the clone A90 into MF1, C57BI/6J, and NMRI

TABLE II
Injection of Clones P4 and A90 into Blastocyst from MF1, C57BI/6J and NMRI Strains[a]

Clone	Host	Blastocyst transferred	Number born	Chimera M	Chimera F	Perinatal deaths	Fertile males	No. progeny
P4	MF1	116	31	3	5	n.d.	3	104
	C57BI/6J	565	21	3	1	n.d.	2	34
A90	MF1	33	4	3	0	n.d.	2	108
	C57BI/6J	463	34	1	4	n.d.	1	59
	NMRI	79	20	8	4	4	5	271
		1256	110	18	14	4	13	576

[a] M, male; F, female; n.d., not determined.

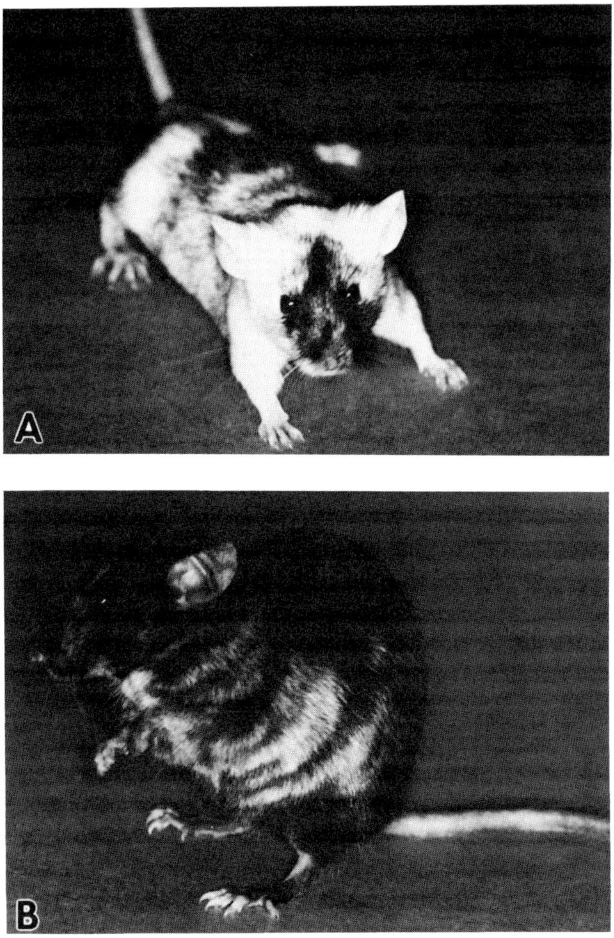

Fig. 4. Chimeric mice generated with the clone A90. (A) A NMRI/A90 chimera. (B) A C57BL/6J chimera. See text for details.

blastocysts. From 575 retransferred blastocysts, we obtained 12 chimeric males. The degree of ES cell contribution to the coat ranged from 10 to 40%. Eight of these males were fertile and produced a total of 438 offspring. None of these offspring was derived from the embryonic stem cells.

DISCUSSION

We have summarized in this report experiments that demonstrate that genes can be mutated at a high frequency in embryonic stem cells without the need for selection. The high frequency of homologous recombination that we observed was very surprising. We calculated that about 25% of all microinjected cells stably integrated the foreign DNA (Capecchi, 1980). Thus, the frequency of homologous recombination compared to illegitimate recombination was 1:30 to 1:50, depending on the type of construct. This is about an order of magnitude greater than the frequency in other reports (Smithies *et al.*, 1985; Thomas *et al.*, 1986; Thomas and Capecchi, 1987; Joyner *et al.*, 1989). The reason for this difference is unclear. Since the screening by PCR can potentially detect any homologous recombination event, our analysis could have simply been more sensitive than those depending on the expression of a marker gene. Alternatively, some chromosomal regions might be more recombinogenic than others.

An answer to this question clearly requires a better understanding of the molecular mechanism involved in homologous and illegitimate recombination, especially determination of which are the rate-limiting steps. There is now accumulating evidence that the search for the homologous chromosomal sequences is not rate limiting. This evidence is based on observations that neither the copy number of target molecules nor the concentration of exogenous DNA affects the frequency of homologous recombination significantly (Thomas *et al.*, 1986; Zheng and Wilson, 1990). If this argument is correct, the steps before or after the formation of a recombination intermediate must be limiting. To clarify this issue, it is necessary to measure the rate with which recombination intermediates are formed. An observation that foreign DNA can induce mutations into target sequences, called *heteroduplex induced mutagenesis* (Thomas and Capecchi, 1986), provides indirect evidence that intermediates are more frequently formed than we can actually observe by homologous recombination. Thus, it is conceivable that the very short nonhomologous sequence used in our construct favored the expected resolution of the recombination intermediate.

Although a better understanding of the recombination process would certainly improve the gene targeting, our method should be generally applicable for the introduction of small mutations, even single-base pair exchanges into ES cells. Thus, we now have the tools for a systematic mutational analysis of gene function in a multicellular organism.

Puzzling still is the lack of germline contribution of the mutated ES cells in the chimeric males. Since *Hox-1.1* is expressed in the testis, we cannot exclude the possibility that the mutation interferes with spermatogenesis.

The rate at which we obtained live offspring was strongly influenced by the genetic background of the host blastocysts and was 5.4% with C57BI/6J, 23.5% with MF1, and 25.3% with NMRI mice. Interestingly, the yield of chimeras was lowest with C57BI/6J mice. Blastocysts of this strain are widely used as recipients, because it has been found that the D3 cell line generally contributes very well to all tissues in those chimeras.

We therefore suspect that initially the ES cells contributed extensively to the developing embryos in C57Bl/6J mice, with the consequence that most of those chimeras died prenatally. In concordance with this assumption is our observation that most foster mothers appeared as if they were carrying large litters, but when they delivered, only a small number or no live offspring were found. It remains to be determined whether this phenomenon is correlated with the mutation in the *Hox-1.1* gene or caused by another mutation in the cell lines used in our experiments.

ACKNOWLEDGMENTS

We thank Uli Eisel and Ted Ustin for carefully reading the manuscript. This work was supported by the Max-Planck Society.

REFERENCES

Balling, R., Mutter, G., Gruss, P., and Kessel, M. (1989). *Cell* **55,** 337–347.

Bradley, A., Evans, M., Kaufmann, M. H., and Robertson, E. (1984). *Nature (London)* **309,** 255–256.

Capecchi, M. R. (1980). *Cell* **22,** 479–488.

Capecchi, M. R. (1989). *Science* **244,** 1288–1292.

Duboule, D., and Dolle, P. (1989). *EMBO J.* **8,** 1497–1505.

Doetschmann, T. C., Eistetter, H., Katz, M., Schmidt, W., and Kemler, R. (1985). *J. Embryol. Exp. Morphol.* **87,** 27–45.

Doetschmann, T., Gregg, R. G., Maeda, N., Hooper, M. L., Melton, D. W., Thompson, S., and Smithies, O. (1987). *Nature (London)* **330,** 576–578.

Evans, M. J., and Kaufmann, M. H. (1981). *Nature (London)* **292,** 154–165.

Gaunt, S. J., Sharpe, P. T., and Duboule, D. (1988). *Development* **104** (Suppl.), 169–179.

Gossler, A., Doetschmann, T., Korn, R., Serfling, E., and Kemler, R. (1986). *Proc. Natl. Acad. Sci. U.S.A.* **83,** 9065–9069.

Graham, A., Papalopulu, N., and Krumlauf, R. (1989). *Cell* **57,** 367–378.

Hooper, M., Hardy, K., Handyside, A., Hunter, S., and Monk, M. (1987). *Nature (London)* **326,** 292–295.

Jasin, M., and Berg, P. (1988). *Genes Dev.* **2,** 1353–1363.

Joyner, A. L., Skarnes, W. C., and Rossant, J. (1989). *Nature (London)* **338,** 153–156.

Kessel, M., Balling, R., and Gruss, P. (1990). *Cell* **61,** 301–308.

Kuehn, M. R., Bradley, A., Robertson, E. J., and Evans, M. J. (1987). *Nature (London)* **326,** 295–298.

Mansour, S. L., Thomas, K. R., and Capecchi, M. R. (1988). *Nature (London)* **336,** 348–352.

Martin, G. R. (1981). *Proc. Natl. Acad. Sci. U.S.A.* **78,** 7634–7638.

Robertson, E., Bradley, A., Kuehn, M. R., and Evans, M. J. (1986). *Nature (London)* **323,** 445–448.

Smithies, O., Gregg, R. G., Boggs, S. S., Koralewski, M. A., and Kucherlapati, R. S. (1985). *Nature (London)* **317,** 230–234.

Thomas, K. R., and Capecchi, M. R. (1986). *Nature (London)* **324,** 34–38.

Thomas, K. R., and Capecchi, M. R. (1987). *Cell* **51,** 503–512.

Thomas, K. R., Folger, K. R., and Capecchi, M. R. (1986). *Cell* **44,** 419–428.

Wolgemuth, D. J., Behringer, R. R., Mosteller, M. P., Brinster, R. L., and Palmiter, R. D. (1989). *Nature (London)* **337,** 461–464.

Zimmer, A., and Gruss, P. (1989). *Nature (London)* **338,** 150–153.

Zheng, H., and Wilon, J. H. (1990). *Nature (London)* **344,** 170–173.

4

Identification and Targeted Mutation of Developmental Genes in Mouse Embryonic Stem Cells

J. ROSSANT, A. GOSSLER, C. MOENS, W. C. SKARNES,
AND A. JOYNER
Division of Molecular and Developmental Biology
Samuel Lunenfeld Research Institute
Mount Sinai Hospital
Toronto, Ontario M5G 1X5
and
Department of Medical Genetics
University of Toronto
Toronto, Ontario M5S 1A1

INTRODUCTION

Our understanding of the genetic control of mammalian embryonic development is still rudimentary when compared that of with *Drosophila* and *Caenorhabditis elegans*. The repository of spontaneous and induced mutations in the mouse has provided some examples of mutations that affect developmental processes (1), but few of these have been molecularly characterized. Serendipitous production of insertional mutations in transgenic mice is adding to the pool of embryonic lethal mutations available (2), and these mutations may be more readily cloned because of the inserted transgene tag. However, it is apparent that the existing mutations are too few in number to provide us with clues as to the overall genetic strategy controlling development of the mammalian embryo. By analogy with other species, it is reasonable to believe that there is a hierarchy of genes controlling lineage specification and also regional specialization within the embryo. The possible nature of some of these genes is beginning to become apparent as molecular cloning of putative mammalian regulatory genes proceeds apace.

Regulation of gene expression in higher organisms acts at various levels within the cell, beginning with signal-transduction pathways converting extracellular signals into intracellular events, and proceeding to the binding of specific transcription factors to the DNA. Genes potentially acting at different parts along this pathway have been postulated as important regulators of embryonic development. The most extensively studied have been the homeobox-containing genes, which contain a conserved DNA-binding domain that is also present in a variety of *Drosophila* genes that control pattern formation (3). There is increasing evidence that at least some of these genes also may be involved in regional specialization in the mammalian embryo, based mostly on their localized pattern of expression during development (4,5,6). Several other classes of genes acting at various points in the signal-transduction pathway are also thought to play roles in embryonic specification. For example, growth factors and growth factorlike molecules seem to be important in induction of mesoderm, at least in amphibian embryos (7,11).

The catalog of putative regulatory genes that have been cloned is increasing in size daily, and the domains of expression of such genes in the developing embryo are being delineated by *in situ* hybridization to messenger RNA (mRNA) and immunocytochemical localization of protein. However, the complete inventory of genes involved in embryonic development has by no means been amassed, and it remains important to identify more genes potentially involved in embryonic lineage formation and pattern development. No genes have been implicated as yet in the first lineage determination events in the mouse embryo, namely formation of the extraembryonic lineages, trophectoderm, and primitive endoderm. Regional specification within the embryo itself is thought to begin around the time of gastrulation and primitive streak formation (8), but few of the homeobox-containing genes have been shown to be regionally localized this early in development. As well as identifying more putative controlling genes, particularly at these early stages of development, it is also critical to be able to test directly the function of any such gene in development. Although various strategies have been devised to test function based on manipulating gene expression in transgenic mice, the most direct test is to generate mutations in the endogenous gene of interest and examine their phenotypes. Mutational analysis will also be required to elucidate the hierarchy of genes involved in any developmental process.

Our recent studies have centered around the use of embryonic stem (ES) cells as tools both for identifying new genes potentially important in development and for generating targeted mutations in any cloned gene of interest.

EMBRYONIC STEM CELL TECHNOLOGY

Embryonic stem (ES) cells are pluripotent cell lines, derived from early embryos, that can be maintained in the undifferentiated state in culture (9,10). On return to the blastocyst environment, however, these cells can display full embryonic potential, contributing to all somatic tissues of resulting chimeras, and most important, to the germ line (11,12,13,14). It is thus possible to genetically manipulate the ES cells in culture by various means and incorporate the genetic alteration produced in culture into the mouse gene pool (Fig. 1). We have used the ability of ES cells to contribute widely to somatic and germ cells in chimeras to devise means of identifying and mutating genes of potential developmental interest.

IDENTIFICATION OF NEW DOMAINS OF GENE ACTION IN DEVELOPING EMBRYOS

The *Escherichia coli* β-galactosidase gene, when driven by a minimal promoter, can be activated by *cis*-acting elements in the host genome, and activity can be detected *in situ* by histochemical staining. Recent studies in *Drosophila* have utilized this to demonstrate novel patterns of reporter gene activation in a large percentage of P element–*lacZ* integration events (15). It is presumed that, in many cases, endogenous genes close to the site of integration show the same or similar patterns of activity and that the integrated vector will allow cloning of these genes. As well as providing a novel route to cloning new developmentally regulated genes, the patterns of reporter gene activation alone provide important new lineage markers for developmental studies.

Reports from our laboratory (16) and that of Surani (17) indicated that similar activation of minimal promoter–*lacZ* constructs could be detected at fairly high frequency in transgenic mice. This suggested to us that it might be possible to use this approach on a larger scale to screen for patterns of reporter gene activation at specific stages of development and thus identify new developmental genes and domains of gene action. To this end we decided to introduce reporter constructs into embryonic stem cells and examine patterns of gene activation in chimeras, thus avoiding the high expenditure of time and money of generating multiple transgenic lines. As reported

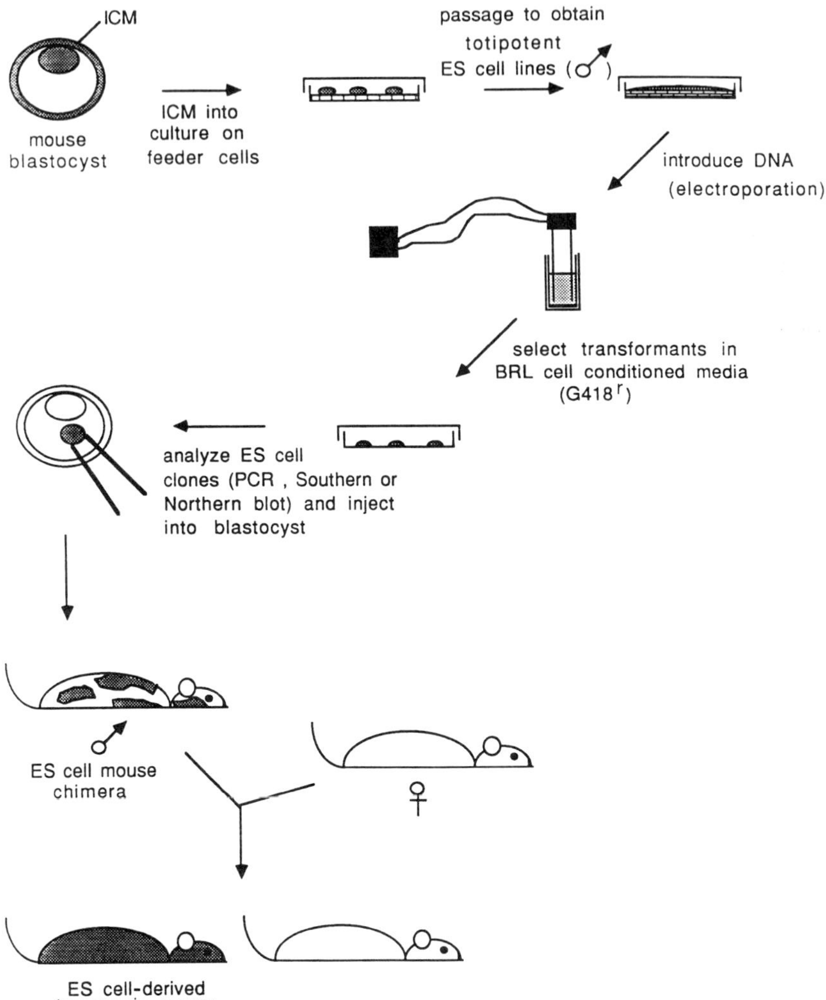

Fig. 1. Strategy for introduction of genetic modifications into the mouse germ line using ES cells.

(18), we have electroporated both enhancer-trap and gene-trap vectors into ES cells and identified examples of localized patterns of *lacZ* expression in the developing chimeric embryos.

In the enhancer trap vector, the *lacZ* gene was fused to a minimal promoter derived from the mouse *hsp68* gene. The vector also contained the neomycin resistance (neo^r) gene driven by the promoter to allow for selection of cells that had taken up DNA. The gene-trap vector contained a promoterless *lacZ* gene lacking a translational initiation signal but containing a splice acceptor signal 5′ to *lacZ*. When this vector integrates in the correct orientation within an intron of a gene, a spliced *lacZ* fusion transcript should be generated, and a functional fusion protein, when the reading frame is maintained. Both constructs, when introduced into ES cells, generated blue colonies but, as expected, the frequency with the gene-trap vector was lower than with the enhancer-trap vector (1 of 60 versus 1 of 7 neo^r colonies). For this reason, both expressing and nonexpressing enhancer-trap lines were tested for *lacZ* activity in chimeras, but only expressing gene-trap lines were tested. However, in both cases, a variety of patterns of *lacZ* activity were observed in chimeras, including very well defined spatially restricted patterns of expression in two of seven enhancer-trap and two of eight gene-trap cell lines. Confirmation that the restricted patterns seen in the chimeras do not reflect some anomalous distribution of the ES cells was obtained when one reporter gene integration was transmitted through the germ line, and embryos hemizygous for the insertion event showed identical but nonmosaic expression patterns to the original ES chimeras.

The high frequency of interesting *lacZ* activation patterns generated in this pilot project suggests that this is a viable means of identifying new domains of gene action. It should also be possible to clone out associated endogenous genes, and in some cases, mutations will also be generated by the insertion events. For the most interesting enhancer-trap insertion event, D3-6-28, which shows expression restricted to the posterior neuropore at day 8.5 of development, we have cloned out flanking sequences to the insertion using the polymerase chain reaction [PCR;(19)] and identified at least one unique sequence probe, which we are currently using to search for putative transcripts. We have also bred further the transgenic offspring carrying this insertion and shown that homozygous offspring can be produced that show no obvious phenotype, suggesting that the particular insertion has not directly disrupted a host gene. Short flanking exon sequences to new gene-trap insertions have been cloned using PCR with complementary DNAs (cDNAs) obtained from the ES cell lines, and isolation of full-length cDNAs from the wild-type genes is underway. Sequencing has confirmed the proper use of the splice acceptor and presence of in-frame open reading frames upstream of *lacZ*. None of these insertions has yet been transmitted to the germ line, but we expect all of these to be mutations, because the coding sequence of the endogenous gene is interrupted by the *lacZ* gene.

We have not proven (June 1989) that the patterns of *lacZ* expression reflect patterns of endogenous gene activity, but this information will soon be available. Even if this is not always the case, the patterns observed may provide useful markers of domains in the developing embryo. It is clear that the gene-trap vector has several advantages. Most important are the ease of cloning the host gene and the inevitable generation of mutations by the insertion event. However, the frequency of activation of *lacZ* is low, making it not really feasible to screen all nonexpressing ES colonies for those that may be activated later in development. Thus, at the moment, the gene-trap approach is realistically restricted to those integration events that activate *lacZ* in ES cells in culture, some of which will be active in restricted patterns in later development. The enhancer-trap approach allows examination of both expressing and nonexpressing ES cell lines, and thus potentially allows identification of genes that are activated later in development. However, cloning of associated genes will be harder, because the integration event need not necessarily be close to coding sequences. Further refinements of both vectors may improve both the frequency of activation and the ease of cloning.

TARGETED MUTATION OF CLONED GENES

The approach outlined above potentially identifies new genes of developmental interest and in some cases mutates them at the same time. However, for insertions, such as D3-6-28, which do not cause mutations, and for any other cloned genes identified as of potential developmental relevance by any other route, technology for generating targeted mutations that can be transmitted into the mouse germ line is essential. As described elsewhere in this volume in greater detail, strategies for achieving targeted mutagenesis by homologous recombination in mammalian cells have recently advanced significantly, such that generation and identification of targeted integration events in most cloned genes of interest is likely to be possible, given suitable targeting vectors. The combination of gene targeting by homologous recombination with ES cell technology thus provides the possibility of introducing defined mutations into the mouse germ line. For mouse developmental biologists, this provides the vital missing tool in the genetic analysis of development.

Because of the importance of this technology for our studies, we have placed considerable emphasis on perfecting the various elements of this strategy in our laboratories. To date, we have generated ES cell lines with targeted mutations in two genes: the mouse engrailedlike gene, *En-2*, whose expression pattern predicts a possible role in regional specialization in the brain (20), and the protooncogene, N-*myc*, in which limited *in situ* analysis suggests a role in differentiation of several lineages (21). In both cases, we have chosen as recipient ES cell line the male D3 129/J-derived line, which has been shown to contribute to the germ line after genetic manipulation and selection in several laboratories (11,18,22). Choice of a male cell line enhances the chances of detecting germ line transmission, because of the large number of offspring one male can produce, and because XY ↔ XX chimeras are often male in phenotype and will transmit only the XY genotype in sperm.

The *En-2* gene was mutated using a *neo*-expressing replacement-type vector at a frequency of approximately 1 of 300 *neo*^{r} colonies (23). Detection of the correct targeted *neo*^{r} colony was performed by a colony-picking procedure, outlined in Fig. 2, combined with PCR. The critical point about this strategy is that it allows rapid identification of the requisite colonies in a

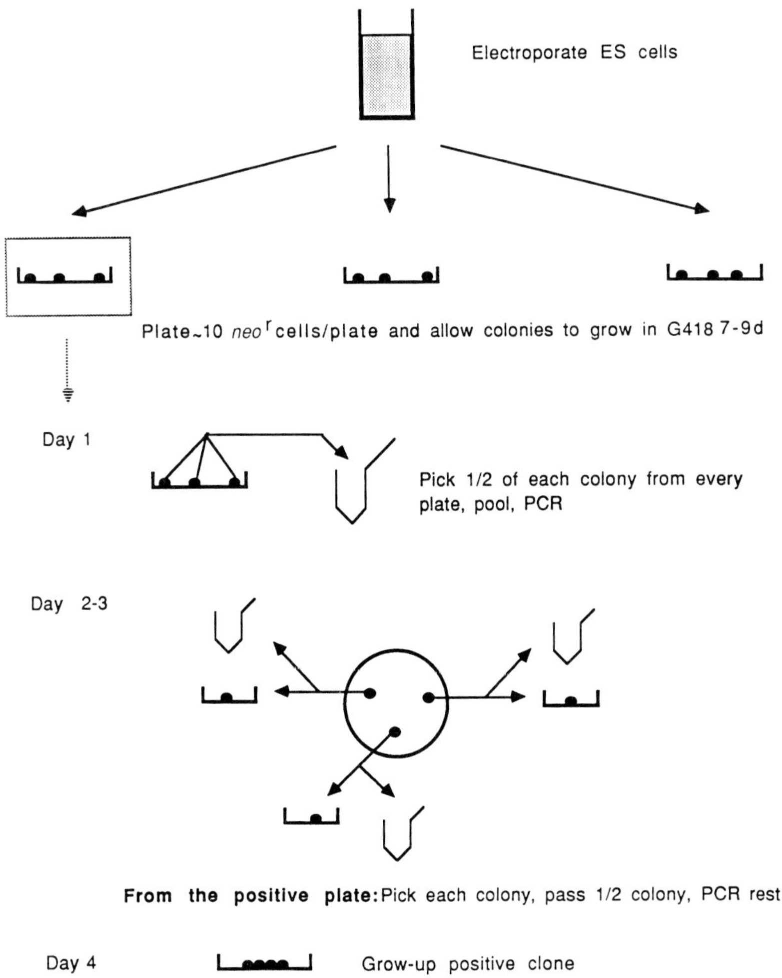

Fig. 2. Strategy for identifying clones of targeted ES cells by PCR and colony-picking.

short period without any necessity for pooling colonies in culture and classic sib selection. By this means we can identify and grow up individual targeted colonies within 2 weeks of initiating the experiment by electroporation. Since it is critical to maintain ES cells in a karyotypically normal state for germ-line transmission, it is desirable to maintain the cells in culture for the minimal length of time before injecting into blastocysts.

The N-*myc* gene is expressed at high levels in ES cells [(24); our unpublished data], allowing one to devise a strategy based on use of a promoterless *neo* construct (Fig. 3). If this vector integrates in the N-*myc* gene by homologous recombination, then an N-*myc neo* fusion transcript should result. Translation should initiate at the start site of the *neo* gene, because there are in-frame stop codons upstream of the *neo* initiation signal. When electroporated into D3 ES cells, *neo*^{r} colonies were obtained with this N-*myc* vector at a frequency 10-fold lower than that with a β-actin–*neo* construct. By using the colony-picking PCR method described in Fig. 2, we were able to show that about 1 of 70 of the *neo*^{r} colonies were targeted integrations in the N-*myc* gene (Table I). Five cell lines have been grown up and confirmed as mutant by Southern blot analysis.

We have shown that three cell lines containing one mutated allele of *En-2*, and one line with an N-*myc* mutation can form chimeras with ES somatic

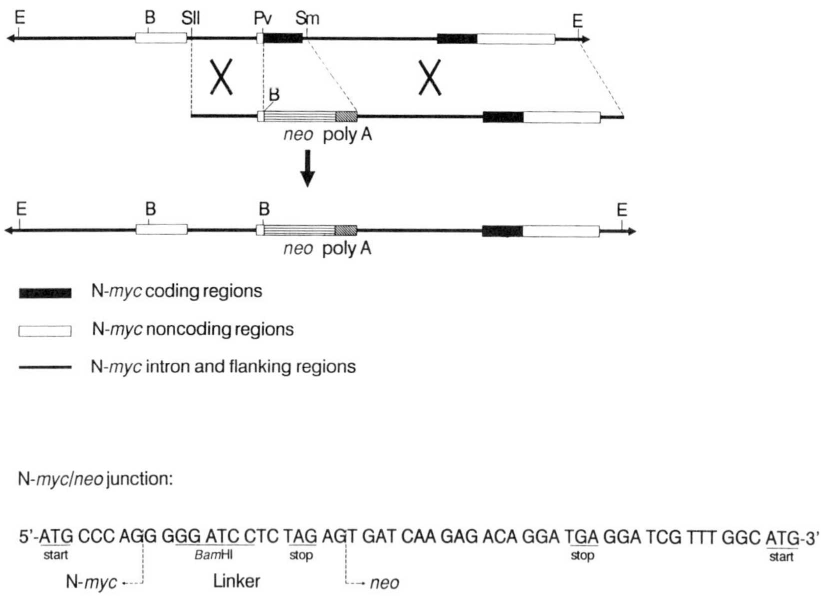

Fig. 3. Scheme for targeting mutations in the N-*myc* gene by homologous recombination.

TABLE I
Frequency of Targeted Integration into N-*myc* Gene in ES Cells

Experiment	Number of cells electroporated	Number of G418^r colonies	Number of homologous recombinants assayed by: PCR	Southern blot
1	5×10^7	100	2	2
2	2.5×10^7	180	3	0
3	7.5×10^7	300	4	3
4	3×10^7	130	1	0
Total	1.8×10^7	710	10	5

contributions. To date we have not achieved germ-line transmission through lack of number of suitable chimeras for test-breeding. Our experience and that of others (C. Stewart, unpublished) has been that the D3 cell line will readily generate chimeras when introduced into blastocysts from outbred albino CD-1 mice. However, extensive somatic tissue contribution is rarely observed, and germ-line transmission of manipulated or unmanipulated D3 cells has not been achieved by us in such chimeras. When the same cells are introduced into inbred embryos, chimeras with much more extensive coat-color contributions result, and germ-line transmission can occur. The necessity of using inbred embryos slows down the rate of generating chimeras, because of the low number of embryos that can be produced, poor quality of the embryos, and lower success of embryo transfer. Nevertheless, we have in hand at least nine male chimeras with extensive N-*myc* mutant ES contribution from injections into inbred blastocysts, two with the ES contribution approaching 100%. An extensive series of injections of the *En-2* mutant cell line is also concluding, and male chimeric offspring have been born. With the numbers of extensive chimeras we have now for test-breeding, we expect to be able to determine whether these mutant cell lines are capable of contribution to germ cells in 1989.

PROBLEMS AND PROSPECTS

The prospects for the application of ES cell technology to identifying and mutating developmental genes seem bright. The problems unresolved in targeted mutagenesis are whether all genes, including nonexpressed genes, can be targeted by homologous recombination and how consistently germ-line transmission can be achieved. On the first issue, it seems clear that application of the double-selection technique devised by Capecchi and coworkers (25) will allow identification of targeted mutations in a much wider spectrum of genes than previously thought feasible. However, this approach still requires some expression of the *neo*^{r} gene in the targeted site. If the frequency of homologous integration of a targeting vector that does not require *neo* selection can be routinely achieved by DNA microinjection at the levels reported by Zimmer and Gruss for the *Hox-1.1* locus (26), then this approach may allow targeting of genes completely silent in ES cells. The second issue of frequency of germ-line transmission will be resolved only when we have more laboratories with more experience in this area. It certainly is possible to introduce homologously recombined (27) and other genetically manipulated ES cells into the germ line; our experience to date suggests that the host strain of the blastocyst may be critical. More experience with different ES cell lines of various genotypes combined with varying host blastocysts is required to optimize the possibility of germ-line transmission. Issues of the effects of the genetic manipulation *per se*, at least as far as *neo*-expressing cells go, seem secondary to the requirements of correct donor–host combinations in the chimeras.

Once these technical issues are resolved, homologous recombination in ES cells will become a vitally important tool for generating all kinds of mutations from "simple" null mutations to subtle mutations in upstream regulatory elements and protein domains. Combined with new approaches to identifying genes involved in developmental processes, this approach offers the possibility for the first time of a detailed analysis of the genetic hierarchy controlling mammalian development.

ACKNOWLEDGMENTS

This work was supported by National Cancer Institute, Medical Research Council, and the Natural Sciences and Engineering Research Council of Canada. J. Rossant is an NCI Scientist; A. Joyner is an MRC Scholar; W. C. Skarnes is supported by an NCI studentship; C. Moens, by NSERC 1967 Scholarship; and A. Gossler, by the Deutsches Forchungs Gemeinschaft.

REFERENCES

1. Green, M. C. (1981). "Genetic Variants and Strains of the Laboratory Mouse." Fischer Verlag, Stuttgart, Germany.
2. Palmiter, R. D., and Brinster, R. L. (1986). *Annu. Rev. Genet.* **20,** 465–499.
3. Gehring, W. J. (1987). *Science* **236,** 1245–1252.
4. Gaunt, S. J. (1988). *Development* **103,** 135–144.
5. Holland, P. W. H., and Hogan, B. L. M. (1988). *Genes Dev.* **2,** 773–782.

6. Graham, A., Papalopulu, N., and Krumlauf, R. (1989). *Cell* **57,** 367–378.
7. Smith, J. C. (1989). *Development* **105,** 665–678.
8. Beddington, R. S. P. (1986). *In* "Experimental Approaches to Mammalian Embryonic Development" (J. Rossant and R. A. Pedersen, eds.), pp. 121–147. Cambridge University Press, New York.
9. Evans, M. J., and Kaufman, M. H. (1981). *Nature (London)* **292,** 154–155.
10. Martin, G. R. (1981). *Proc. Natl. Acad. Sci. U.S.A.* **78,** 7634–7638.
11. Gossler, A., Doetschman, T., Korn, A., Serfling, E., and Kemler, R. (1986). *Proc. Natl. Acad. Sci. U.S.A.* **83,** 9065–9069.
12. Robertson, E., Bradley, A., Kuehn, M., and Evans, M. J. (1986). *Nature (London)* **323,** 445–448.
13. Hooper, M., Hardy, K., Handyside, A., Hunter, S., and Monk, M. (1987). *Nature (London)* **326,** 292–295.
14. Kuehn, M. R., Bradley, A., Robertson, E. J., and Monk, M. (1987). *Nature (London)* **326,** 295–298.
15. O'Kane, C. J., and Gehring, W. J. (1987). *Proc. Natl. Acad. Sci. U.S.A.* **84,** 9123–9127.
16. Kothary, R., Clapoff, S., Brown, A., Campbell, R., Peterson, A., and Rossant, J. (1988). *Nature (London)* **335,** 435–437.
17. Allen, N. D., Cran, D. G., Barton, S. C., Hettle, S., Reik, W., and Surani, M. A. (1988). *Nature (London)* **333,** 852–855.
18. Gossler, A., Joyner, A. L., Rossant, J., and Skarnes, W. C. (1989). *Science* **244,** 463–465.
19. Saiki, R. K., Gelfand, D. H., Stoffel, S., Scharf, S. J., Higuchi, R., Horn, G. T., Mullis, K. B., and Erlich, H. A. (1988). *Science* **239,** 487–491.
20. Davis, C. A., Noble-Topham, S. E., Rossant, J., and Joyner, A. L. (1988). *Genes Dev.* **2,** 361–367.
21. Mugraer, G., Ale, F. W., and Ekblom, R. (1988). *J. Cell. Biol.* **107,** 1325–1335.
22. Williams, R. L., Milton, D. J., Pease, S., Willson, T. A., Stewart, C. L., Gearing, D. P., Wagner, E. F., Metcalfe, D., Nicola, N. A., and Gough, N. M. (1988). *Nature (London)* **336,** 684–687.
23. Joyner, A. L., Skarnes, W. C., and Rossant, J. (1989). *Nature (London)* **338,** 153–156.
24. Jakobovits, A., Schwab, M., Bishop, J. M., and Martin, G. R. (1985). *Nature (London)* **318,** 188–191.
25. Mansour, S. L., Thomas, K. R., and Capecchi, M. R. (1988). *Nature (London)* **336,** 489–492.
26. Zimmer, A., and Gruss, P. (1989). *Nature (London)* **338,** 151–153.
27. Thompson, S., Clarke, A. R., Pow, A. M., Hooper, M. L., and Melton, D. W. (1989). *Cell* **56,** 313–321.

PART II

DNA INSERTION PHENOMENA

5

Developmental Mutations Generated by Retroviral Insertional Mutagenesis

DOUGLAS A. GRAY[1], HANS WEIHER[2], THOMAS GRIDLEY[3], TETSUO NODA[4], ARLENE SHARPE, AND RUDOLF JAENISCH

Whitehead Institute for Biomedical Research
Cambridge, Massachusetts 02142

INTRODUCTION

Baroque though they may seem, the developmental programs of such metazoans as *Drosophila melanogaster* and *Caenorhabditis elegans* are yielding to elegant experiments made possible by the great number of developmental mutants available in these systems. Because the developing mouse is at once more complex and more difficult to manipulate in its uterine environment, our understanding of murine development has lagged behind that of the invertebrate systems. The problem is made much worse, however, by the relative dearth of mutant mouse strains. As a historical consequence of breeding coat variations of mice (the so-called "mouse fancy"), a number of mouse strains have been identified with mutations in hair production and coloration. Two such mutations (*dilute* and *hairless*) have been shown to be the result of retroviral insertion events (1,2), providing a means of locating and molecularly cloning the mutated gene, using the provirus as a molecular "tag." It has been shown (3) that spontaneous germ-line infections could be manipulated to produce an early embryonic lethal mutation (designated Srev-5), although the identity of the mutated gene is unknown. Other

1. *Present address:* University of Ottawa, Ottawa, Canada K1H8M5
2. *Present address:* German Cancer Research Center, D-6900 Heidelberg 1, Germany
3. *Present address:* Roche Institute for Molecular Biology, Nutley, New Jersey
4. *Present address:* Cancer Research Institute, 1-37-1 Kamiikebukuro, Toshima-Ku, Tokyo, Japan

novel mutations have been generated by microinjection of DNA directly into zygotes (reviewed and referenced in (4)]. All of these mutations arise from insertion of foreign DNA into some important genomic site (insertional mutagenesis). Our efforts have been concentrated on providing new developmental mutations in the mouse through breeding to homozygosity transgenic mice, resulting from infection of embryos with retroviruses. To date, four recessive lethal mutations have been identified in our laboratory; these manifest their effect at quite different stages of development. These mutations are being analyzed to determine what information they can add to what is known of the mammalian developmental program. What follows is a brief description of the phenotype of these mutations, and what is known so far of the molecular mechanisms underlying these phenotypes.

RESULTS AND DISCUSSION

Of 70 strains bred to homozygosity, four were found to have recessive lethal mutations. The identity of only one of the four mutated genes is known at the time of writing. In the Mov13 strain, a Moloney murine leukemia virus (M-MuLV) provirus has integrated in the first intron of the α1(I) collagen gene (5,6). The effect of this provirus on transcription of this gene (7) and the resulting death of homozygotes at the twelfth day of gestation from the rupture of major vessels (8) is well documented, and will not be discussed further.

A second mutation (Mov34) was found to arise from the integration of an M-MuLV-derived provirus into an as yet unidentified gene (9). Embryos homozygous for the Mov34 integration were found to develop normally to blastocysts, then die after implanting into the uterine wall. It has previously been reported (9) that the proviral integration suppresses transcription from the gene at this locus, a gene that is normally constitutively expressed in all tissues examined. Chromosome mapping by somatic hybrid analysis and *in situ* hybridization suggested that the integration was close to oligosyndactyly (Os) and might even be causing the same mutation, since Os/Os embryos die at about the same stage of development (10), but crosses with OS mice have ruled this out (11).

The striking conservation of the Mov34 gene (9) led us to examine its homolog in an organism whose development is better understood, namely *Drosophila*. Both complementary DNA (cDNA) and genomic clones from *Drosophila* phage libraries were isolated, using a Mov34 cDNA probe (11). Expression in *Drosophila* was also found to be constitutive, allowing detection of a transcript similar in size to that observed in mouse tissues. The nucleotide sequences of mouse and fly cDNAs were determined, and as expected, were quite similar, encoding putative proteins with 60% identity at the amino acid level. *In situ* hybridization of the *Drosophila* probes to polytene chromosomes places the Mov34 homolog in the vicinity of several known mutations, some of which affect larval development (11). None of the sequences obtained to date have significant similarity to sequences already in the databases. Although the putative Mov34 proteins from mouse and fly both have potential transmembrane domains and extremely hydrophilic carboxy termini, nothing more can be inferred of their function, except to say their ubiquity and conservation suggest a "housekeeping" role, which (at least in the mouse) is provided maternally or is not required for a number of cell divisions in embryo.

More recently, mutations have been generated through infection of embryos with a replication-defective derivative of myeloproliferative sarcoma virus (MPSV). Mice homozygous for the Mpv17 provirus die as adults of nephrotic syndrome leading to kidney failure (12). Histologic examination of tissues from affected mice revealed that illness was owing to a glomerular disease resembling mesangioproliferative glomerulonephritis. Changes were noted in serum, blood, and urine values consistent with nephrotic syndrome and chronic renal failure. Once again, molecular cloning of the integration site revealed that the provirus had integrated within or adjacent to a gene whose expression was adversely affected by its presence. This gene was found to be conserved in vertebrates as distantly related as fish. Nucleotide sequence of a kidney cDNA isolate revealed an open reading frame sufficient to encode a small protein. The identity of this gene and its protein remains unknown, and whereas its ubiquitous expression suggests a housekeeping function, the tissue specificity of the mutation and its very late onset argue strongly for a specific function in adult kidneys.

A fourth mutation (Mpv20) was also generated by infection of embryos with the MPSV-derived virus. This mutation represents the earliest known embryonic lethal mutation. Homozygous Mpv20 mice apparently die as uncompacted morulas of four to eight cells (13). Molecular cloning of the Mpv20 integration site allowed for the identification of a region conserved in mammalian species, although apparently not beyond. A genomic probe from this region detected a transcript in developing embryos up to the time of birth, as well as in embryonic carcinoma and stem-cell lines. Quite different-sized transcripts were seen in adult tissues, each tissue having its unique pattern of expression. cDNA libraries have now been prepared from some of these sources (both embryonic and adult), and hybridizing clones have been isolated. A detailed examination of these clones is under way.

SUMMARY AND CONCLUSION

The efficiency of generating recessive mutations by retrovirus infection of embryos appears to be about 6% (4 mutants from 70 insertion events). The

TABLE I
Virus-Induced Insertional Mutations

Locus	Homozygous phenotype (time of death)	Time of activation	Tissue distribution	Gene (RNA size)
MPV20	Preimplantation (8-cell stage)	Early (F9, ES)	Many tissues	?1.5 kilobases (kb) in embyros; tissue-specific in adults
Mov34	Day 5 gestation (postimplantation)	Early (F9, ES)	Most or all tissues	? (1.7 kb)
Mov13	Day 13 gestation	Day 11 gestation	Fibroblasts, mesenchymal cells	COL1A1 [α1(I) collagen]
MPV17	Adult (kidney failure)	Early (F9, ES)	Many tissues	? (1.8 kb)

method, therefore, entails the use of a considerable number of mice to generate even a few mutations and would represent a high-risk experiment were it not for the nature of the mutations so generated. Even though there is currently no way to select for interesting developmental mutations (avoiding those housekeeping genes whose mutation should result in cell-lethal effects), it is clear that insertional mutagenesis does generate mutations that are distributed along the developmental pathway of the mouse. If one considers the data from the above-mentioned four mutant strains (summarized in Table I), one can see that the effects of these different mutations are manifested essentially over the lifetime of the mouse, from the very early embryo (Mpv20), to the implantation stage embryo (Mov34), to the midgestation embryo (Mov13), and on to the adult mouse (Mpv17). The use of the provirus as a molecular "tag" to locate and isolate the affected gene has provided us with the means to study the activation and regulation of these genes. Often, we are left with interesting puzzles. Why, for example, is the effect of the Mpv17 mutation limited to the kidney, when transcription of the gene is disrupted in all tissues? How are families of Mpv20 transcripts generated in adult tissues, when only a single transcript is detected in embryonic tissues? In addition to determining the identity of the genes we have mutated, we are currently striving to answer these questions.

REFERENCES

1. Jenkins, N., Copeland, N., Taylor, B., and Lee, B. (1981). *Nature (London)* **293**, 370–374.
2. Stoye, J., Fenner, S., Greenoak, G., Moran, C., and Coffin, J. (1988). *Cell* **54**, 383–391.

3. Spence, S., Gilbert, D., Swing, D., Copeland, N., and Jenkins, N. (1989). *Mol. Cell Biol.* **9,** 177–184.
4. Jaenisch, R. (1988). *Science* **240,** 1468–1471.
5. Schnieke, A., Harbers, K., and Jaenisch, R. (1983). *Nature (London)* **304,** 315–320.
6. Harbers, K., Kuehn, M., Delius, H., and Jaenisch, J. (1984). *Proc. Natl. Sci. U.S.A.* **81,** 1504–1508.
7. Hartung, S., Jaenisch, R., and Breindl, M. (1986). *Nature (London)* **320,** 365–367.
8. Löhler, J., Timpl, R., and Jaenisch, R. (1984). *Cell* **38,** 597–607.
9. Soriano, P., Gridley, T., and Jaenisch, R. (1987). *Genes Dev.* **1,** 366–375.
10. Van Valen, P. (1966). *J. Embryol. Exp. Morphol.* **15,** 119–124.
11. Gridley, T., Gray, D., Orr-Weaver, J., Soriano, P., Barton, D., Francke, U., and Jaenisch, R. In preparation.
12. Weiher, H., Noda, T., Gray, D., Sharpe, A., and Jaenisch, J. In preparation.
13. Gray, D. Unpublished data.

6

Transfer of Yeast Artificial Chromosomes into Cultured Cells: A New Method for Manipulating the Mammalian Genome

VASSILIS PACHNIS, LARYSA PEVNY,
RODNEY ROTHSTEIN, AND FRANK COSTANTINI
Department of Genetics and Development
Columbia University
New York, New York 10032

INTRODUCTION

The ability to transfer cloned genes into mammalian cells, via methods such as calcium phosphate precipitation, protoplast fusion, electroporation, or microinjection, has been an invaluable tool for studying mechanisms of gene regulation, as well as for investigating the functions of specific gene products in cellular differentiation and physiology. One limitation on the application of this technique has been that the size of a cloned DNA segment to be transferred could not exceed approximately 50 kilobases (kb), the capacity of a bacterial cosmid vector. Recently, however, yeast artificial chromosomes (YACs) have been developed as cloning vehicles for the propagation in *Saccharomyces cerevisiae* of very large segments of chromosomal DNA, hundreds of kilobases in length (1). The ability to introduce intact YACs back into mammalian cells, and potentially into embryos, would permit cellular transformation with large genes or gene complexes, as well as the screening of large segments of the mammalian genome to identify genes capable of complementing recessive mutations, or conferring dominant mutant phenotypes.

We have examined the feasibility of transferring YAC DNA from *S. cerevisiae* into murine fibroblasts, using cell-fusion techniques that had been previously used to transfer large segments of chromosomal DNA (or in some cases entire chromosomes) from one mammalian cell to another. Our studies have demonstrated that a YAC, carrying approximately 450 kb of human DNA as well as a neomycin resistance gene, could be transferred efficiently from the *S. cerevisiae* cells into the mouse cells, where it was stably integrated into the host chromosomes (8).

RESULTS

We employed in these experiments the YAC "HY19," which carried approximately 450 kb of human DNA from an unidentified locus. To permit drug selection of any L cells acquiring the human DNA sequences, we first modified the HY19 YAC by inserting the MC1POLA neomycin resistance gene (2), as shown in Fig. 1. A 4.0-kb *Sac*I fragment of human DNA, subcloned from the insert of the YAC HY19, was modified by substituting a 3.2-kb *Bam*HI fragment, containing the MC1POLA (*neo*) gene and the *ADE2* gene (a yeast-selectable marker), for the internal 0.6-kb *Bam*HI fragment. AB1380-HY19 cells were transformed (3,4) with the *Sac-ADE2-neo* fragment (Fig. 1, middle). In one out of ten Ade^{+} transformants analyzed (AB1380-HY19-*neo*), the endogenous wild-type *Sac*I fragment had been replaced, by homologous recombination (5), with the modified *Sac-ADE2-neo* fragment.

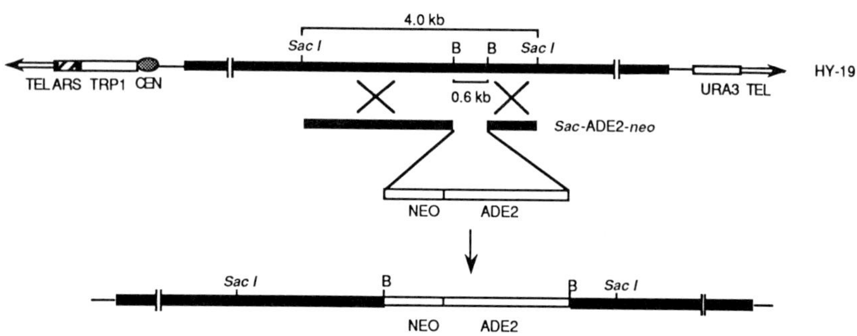

Fig. 1. Insertion of a neomycin resistance gene into the HY19 chromosome by homologous recombination in yeast. Top line, the original HY19 chromosome (not drawn to scale). TEL, telomeres; ARS, yeast autonomous replication sequence; *TRP1* and *URA3;* selectable markers on the HY19 YAC. Middle lines, *Sac-ADE2-neo* fragment containing the selectable markers *ADE2* and *neo*, which was inserted by homologous recombination into HY19. Bottom, part of the modified YAC (HY19-*neo*). B, *Bam*HI. Solid bars indicate human DNA. Reproduced from (8) by permission of the publisher.

Further analysis indicated that AB1380-HY19-*neo* had acquired approximately 20 copies of the *Sac-ADE2-neo* fragment in a head-to-tail arrangement, resulting in a modified HY19-*neo* chromosome somewhat larger (600 kb) than the original HY19 chromosome. The integration of the exogenous sequences on the HY19 chromosome, using the strategy depicted in Fig. 1, established that human DNA propagated in a YAC vector could be specifically modified by homologous recombination in *S. cerevisiae*.

To introduce the modified YAC HY19-*neo* into mouse L cells, we fused the L cells with spheroplasts of the host *S. cerevisiae* strain carrying the YAC (AB1380-HY19-*neo*), and subsequently selected for G418-resistant (G418^r) L cells. This strategy was suggested by the previous demonstration that a yeast plasmid could be transferred into mammalian cells by spheroplast fusion [(6) and our own unpublished data), and by recent studies in which an intact chromosome from the fission yeast *Schizosaccharomyces pombe* had been transferred into mouse cells by similar techniques (7). Our protocol, which was derived from that of Allshire *et al.* (7), has been described in detail (8). Briefly, yeast spheroplasts were generated by digesting the cell wall with zymolyase, then fused to L cells in the presence of polyethylene glycol 1500. The cells were subsequently cultured in the presence of G418, and approximately 50 G418^r L cell colonies were obtained per 10^6 original L cells. The same frequency of G418^r colonies was obtained when L cells were fused with AB1380 yeast containing a multiple-copy plasmid carrying the *neo* gene (unpublished data), indicating that the transfer of YAC sequences was no less efficient than the transfer of plasmid sequences.

To estimate how much of the human DNA from YAC HY19-*neo* had been transferred into the G418^r L cells, DNA from 19 independent clones as well as DNA from the intact HY19-*neo* chromosome was analyzed by restriction enzyme digestion and Southern hybridization, using a human Alu repeat sequence probe (9) (analysis of eight clones is shown in Fig. 2). Sixteen of these L cell clones contained human DNA by this criterion, and nine of the clones (including YF3-6, -10, -26, and -27, in Fig. 2) generated a set of bands ("Alu fingerprint") virtually identical to that of HY19-*neo* DNA. Assuming that Alu repeat elements are dispersed throughout the 450 kb of human DNA, this suggested that the majority of the HY19-*neo* chromosome was transferred in these nine clones.

To test whether any gross rearrangements of the human DNA had occurred on transfer into the L cells, we used two rare-cutting restriction endonucleases in pulsed-field gel analysis (10,11). When YAC HY19-*neo* DNA was digested with either *Xho*I or *Cla*I, a single restriction fragment of 350 kb was detected with a *neo* probe. Fragments of the same size were observed using DNA from L cell clones YF3-10, -26, and -27, indicating that no rearrangement had occurred within the 350-kb *Xho*I or *Cla*I fragments

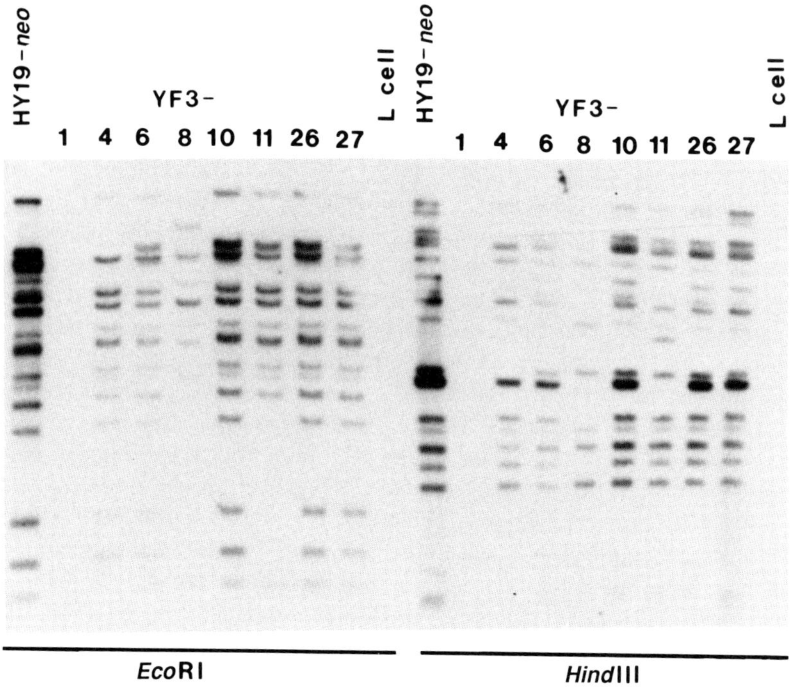

Fig. 2. Analysis of human DNA sequences in L cell transformants by Alu fingerprinting. DNA from AB1380-HY19-*neo* cells (150 ng), the parental L cells (10 μg), and eight YF3 clones (10 μg), was digested with *Eco*RI and *Hin*dIII and electrophoresed on a 1% agarose gel, followed by Southern analysis using a human Alu repeat probe. Similar results were obtained with two additional restriction enzymes. Reproduced from (8) by permission of the publisher.

(Fig. 3). In clone YF3-6, a slightly smaller fragment was observed with both enzymes (Fig. 3); this was the result of excision of several of the tandem *Sac-ADE2-neo* copies by intrachromosomal recombination (data not shown).

To determine whether the entire YAC, including subtelomeric vector sequences, was present in the L cell transformants displaying a complete Alu fingerprint, DNA from YF3-6, -10, -26, and -27 cells was hybridized with URA3 and TRP1 probes that define the left and right YAC arms, respectively (Fig. 1). Clone YF3-6 was positive with both probes, whereas the other three clones were negative with both probes (data not shown). Thus, clone YF3-6 seemed to have acquired the YAC arms as well as the entire human DNA segment present on the HY19 chromosome, whereas clones YF3-10, -26, and -27 had acquired most of the human DNA but lost both ends of the YAC.

Allshire *et al.* (7) have reported that an intact chromosome from *S. pombe* could be maintained as an unstable, extrachromosomal, autonomously rep-

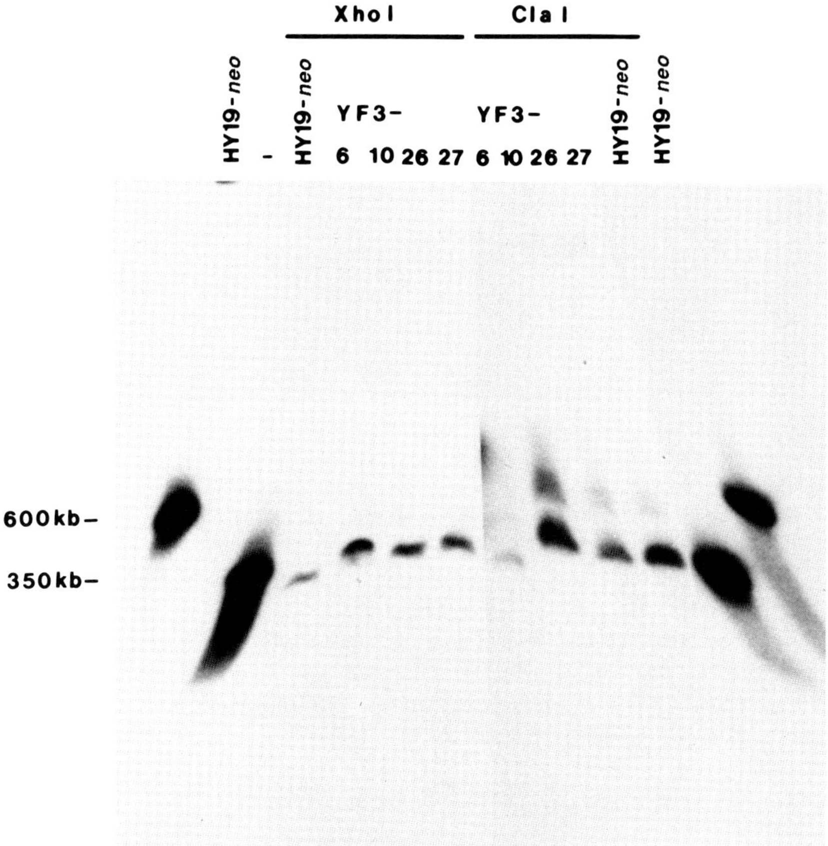

Fig. 3. Pulsed-field gel analysis of the HY19-*neo* DNA in four G418[r] L cell transformants. Undigested high-molecular-weight DNA (16) from AB1380-HY19-*neo* cells (first and last lanes) was electrophoresed (17) in parallel with *Xho*I- or *Cla*I-digested DNA from AB1380-HY19-*neo* and YF3-6, -10, -26, and -27 cells. The DNA was then analyzed by Southern hybridization using a *neo* probe. Reproduced from (8) by permission of the publisher.

licating element in mouse cells. To determine the status of the essentially intact HY19 chromosome in YF3-6 cells, as well as that of the HY19 sequences in the other L cell transformants, metaphase chromosome spreads from clones YF3-6, -26, and -27 were hybridized *in situ* with a probe specific for sequences in the YAC. A single site of hybridization, associated with a unique mouse chromosome, was observed in each cell line examined (e.g., Fig. 4). As an independent test for chromosomal integration of the *neo* gene, the L cell clones were grown in the absence of G418 for seven passages (30–40 generations), and then tested for their relative plating efficiency in the

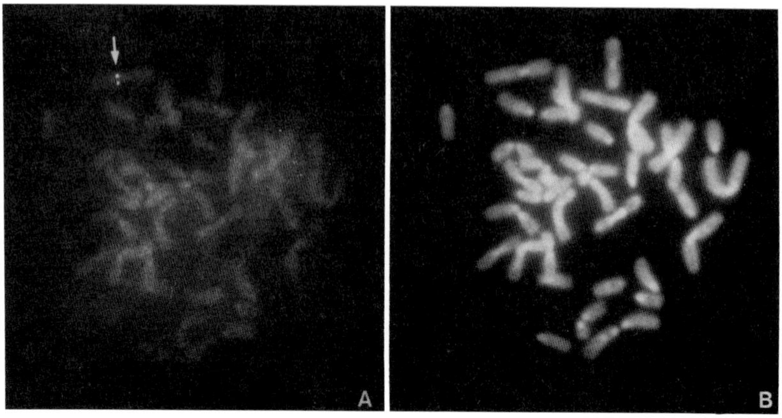

Fig. 4. Detection of YAC DNA inserted into a mouse chromosome, by *in situ* hybridization (6). (A) A representative metaphase spread from YF3-27 cells hybridized with a biotinylated *Sac-ADE2-neo* fragment, which was then visualized with fluorescein isothiocyanate (FITC)-conjugated avidin and fluorescence microscopy. (B) The same metaphase spread counterstained with 4′,6′-diamidino-2-phenylindole (DAPI). Reproduced from (8) by permission of the publisher.

presence or absence of G418. Three of the four clones displayed complete stability of G418 resistance by this assay, consistent with the observed integration of HY19-*neo* sequences into mouse chromosomes.

CONCLUSIONS AND FUTURE PROSPECTS

These studies demonstrated that the fusion of *S. cerevisiae* spheroplasts with mammalian cells results in the efficient transfer of a YAC containing several hundred kb of human DNA from yeast into the mammalian nucleus. In approximately 50% of such transformants, the entire complement of the human cloned sequences was stably integrated into the mouse chromosomes without apparent rearrangement.

The human DNA on the YAC we employed has not been characterized, and therefore we could not determine whether any human genes were expressed after transfer of the YAC into L cells. However, the *neo* gene, which utilizes a mammalian promoter and enhancer (2), was obviously expressed in the L cells, as this was the basis for their selection. Therefore our experiments show that passage of mammalian (YAC) DNA through *S. cerevisiae* does not result in the permanent inactivation of gene expression.

The ability to transfer large continuous segments of mammalian DNA from yeast cells back into mammalian cells should facilitate expression studies with very large genes (e.g., factor VIII or even dystrophin), gene complexes [e.g., immunoglobulin or major histocompatibility complex (MHC)], or genes with distant regulatory elements (e.g., globin genes). In addition, mammalian DNA sequences within the YAC may be mutated while in the yeast host cell, using the powerful genetic techniques available in *S. cerevisiae;* the effects of such mutations could be tested following transfer of the modified YAC DNA back into mammalian cells.

The application of these methods to murine embryonic stem cells (12,13), which efficiently contribute to both somatic and germ-cell lineages when injected into early mouse embryos (14), could permit the generation of transgenic mice carrying any YAC of interest. This would be very useful for studies of developmental gene regulation, or for attempts to screen large regions of mouse or human DNA for genes that can complement existing mouse mutations, or confer dominant mutant or disease phenotypes. To date, however, our attempts to transfer YAC HY19-*neo* into embryonic stem cells, via spheroplast fusion, have not been successful. An alternative route that might permit generation of transgenic mice carrying a large YAC is the direct microinjection of carefully purified YAC DNA (15) into the mouse zygote (J. Gearhart, personal communication).

None of the L cell clones we examined appeared to carry the HY19-*neo* YAC as an autonomously replicating extrachromosomal element, suggesting that structural elements necessary for the function of a mammalian chromosome are missing from the HY19-*neo* YAC. Such a YAC could serve as a framework for the addition and testing of structural elements (e.g., centromeres), derived from the mammalian genome, that could constitute a stable autonomously replicating unit in mammalian cells.

ACKNOWLEDGMENTS

The studies described in this paper have been previously reported (8). This work was supported by grants from the N.I.H. to F. C. (HD17704 and CA22376) and to R. R. (GM34587). V. P. is a Special Fellow of the Leukemia Society of America, L. P. is a predoctoral fellow of the National Foundation—March of Dimes, R. R. is an American Heart Association Established Investigator, and F. C. is a recipient of an American Cancer Society Faculty Research Award.

REFERENCES

1. Burke, D., Carle, G. F., and Olson, M. V. (1987). *Science* **236,** 806–812.
2. Thomas, K. R., and Capecchi, M. R. (1987). *Cell* **51,** 503–512.
3. Sherman, F., Fink, G. R., and Hicks, J. B. (1986). *In "Laboratory Course Manual for Methods in Yeast Genetics."* Cold Spring Harbor Laboratory, New York.

4. Burgers, P. M., and Percival, K. J. (1987). *Anal. Biochem.* **166,** 391–397.
5. Rothstein, R. J. (1983). *Methods Enzymol.* **101,** 202–211.
6. Ward, M., Scott, R. J., Davey, M. R., Clothier, R. H., Cocking, E. C., and Balls, M. (1986). *Somatic Cell Mol. Genet.* **12,** 101–109.
7. Allshire, R. C., Cranston, G., Gosden, J. R., Maule, J. C., Hastie, N. D., and Fantes, P. A. (1987). *Cell* **50,** 391–403.
8. Pachnis, V., Pevny, L., Rothstein, R., and Costantini, F. (1990). *Proc. Natl. Acad. Sci. U.S.A.* **87,** 5109–5113.
9. Jelinek, W. R., Toomey, T. P., Leinwand, L., Duncan, C. H., Biro, P. A., Choudary, P. V., Weissman, S. M., Rubin, C. M., Houck, C. M., Deininger, P. L., and Schmid, C. D. (1980). *Proc. Natl. Acad. Sci. U.S.A.* **77,** 1398–1402.
10. Schwartz, D., and Cantor, C. R. (1984). *Cell* **37,** 67–75.
11. Carle, G. F., and Olson, M. V. (1984). *Nucleic Acids Res.* **12,** 5647–5664.
12. Evans, M. J., and Kaufman, M. H. (1981). *Nature (London)* **292,** 154–156.
13. Martin, G. R. (1981). *Proc. Natl. Acad. Sci. U.S.A.* **78,** 7634–7638.
14. Bradley, A., Evans, M., Kaufman, M. H., and Robertson, E. (1984). *Nature (London)* **309,** 255–256.
15. Couto, L. B., Spangler, E. A., and Rubin, E. M. (1989). *Nucleic Acids Res.* **17,** 8010.
16. Rothstein, R. J. (1985). *In "DNA Cloning: A Practical Approach"* (D. M. Glover, ed.), pp. 45–66. IRL Press.
17. Chu, G., Volrath, D., and Davis, R. W. (1986). *Science* **234,** 1582–1585.
18. Lichter, P., Cremer, T., Borden, J., Manuelidis, L., and Ward, D. C. (1988). *Hum. Genet.* **80,** 224–234.

7

Sex-Dependent *de Novo* Methylation of the Transgene and Its Insertional Locus on Mouse Chromosome 13

X. WU, M. HADCHOUEL, H. FARZA, L. AMAR,
AND C. POURCEL
Unité de Recombinainson et Expression Génétique
28 Rue du Dr. Roux
75724 Paris, Cedex 15
France

INTRODUCTION

Over the years, transgenic mouse strains have been produced with the primary goal of analyzing the expression of the injected DNA. It became rapidly evident that the genomic environment was affecting the expression of the transgene. This was particularly true for genes that were controlled by weak regulatory signals or for some viral sequences such as murine retroviruses of hepatitis B virus (HBV).

One interesting aspect of this position effect is the sex-specific imprinting imposed during passage through the germ line of the animal that leads in some cases to modifications of the transgene expression. We have observed this phenomenon in one strain of transgenic mice containing HBV DNA and expressing the viral surface antigen, but not in a second strain that contains a similar transgene at a different locus. In this second strain, however, the insertion of the viral DNA causes a dominant mutation affecting growth and fertility.

Mechanisms of Eukaryotic DNA Recombination

RESULTS

In order to analyze the tissue-specific and developmental regulation of hepatitis B surface antigen (HBsAg) gene, we produced mice containing the HBV genome deleted of the core gene but containing all the regulatory signals necessary for S gene expression. A 2.8-kb fragment of DNA was cloned in PBR322, and the linearized plasmid was injected into one-cell embryos. Two of six transgenic mice produced HBsAg in the serum. Transcription of the S gene was liver specific and started around day 15 of development, together with that of albumin.[1] A consequence of this was the immunotolerance of the mice toward the HBsAg and the absence of pathology.

In the first strain, E11, progeny receiving the transgene from either a male or a female always expressed HBsAg (Fig. 1). However, there was a very low number of positive animals in each litter of a heterozygote transgenic, especially of females. Many heterozygote fetuses died *in utero*, and homozygotes could not be obtained. We have cloned the transgene and mouse flanking sequences and have mapped the insertion to chromosome 14 next to the glutamate dehydrogenase (*Gdh-x*) gene and Np-1.

In the E36 strain, transmission of the transgene by males was normal and always resulted in HBsAg expression (Fig. 1). Females transmitted the transgene to 50% of their progeny, but none expressed HBsAg. This was not reversible by subsequent passage through the male germ line. Homozygotes were obtained with one or two modified alleles. They developed normally and were fertile.

Restriction map analyses showed that no rearrangement had occurred, suggesting that sequences controlling the S gene expression had been modified. Thus the methylation status of the transgene was investigated in different organs of progeny from males and females, positive or negative for S gene expression (Fig. 2). Three different degrees of methylation were found on restriction with *Hpa*II, which has 24 sites in pBR322 and 3 sites in HBV. In progeny of females, all the sites were methylated in all the tissues tested, including testis and sperm. In progeny of males the transgene was undermethylated in all the organs except for one site in the S gene, which was hypomethylated in the liver only. This site is methylated in sperm and in embryo and becomes progressively demethylated during liver differentiation. On the contrary, the total methylation resulting from passage through the female germ line is not erased during development. In the E11 strain the methylation pattern is similar to that of HBsAg-positive E36 mice, whether the transgene is transmitted by a male or a female.

The DNase I sensitivity of the transgene was then analyzed and two sites were localized in the HBV-enhancer and X-coding region of E11 and E36

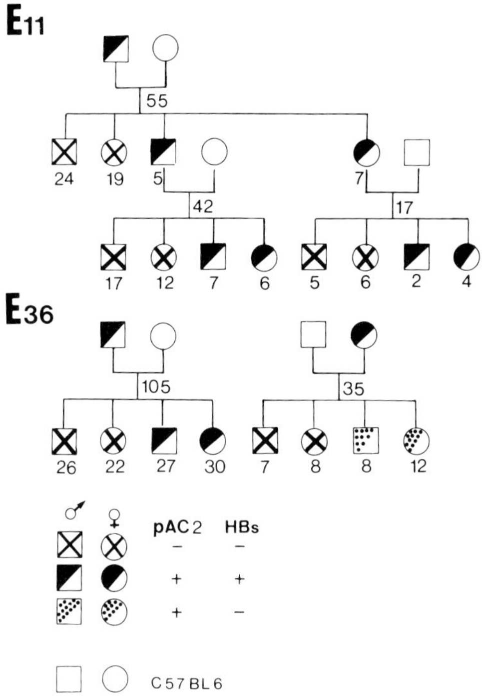

Fig. 1. Pedigree chart of strain E11 and strain E36. Transgenic mice, heterozygote for the transgene, were crossed to C57 BL/6-J-Pas mice. The offspring were tested for HBV DNA using the dot-hybridization technique. Serum HBsAg was measured with the AUSRIA radioimmunoassay (Abbott). The total number of offspring is indicated under each cross, and the number of mice of each type, indicated under the corresponding symbol. Top: E11 founder male produced one offspring by *in vitro* fertilization. This mouse (male F_1) was fertile and produced male and female offspring expressing HBsAg. These were also able to transmit the transgene, although to only about 20% of their progeny. Bottom: E36 founder male produced 50% male and female offspring containing HBV DNA and producing serum HBsAg. Progeny of a male always expresses the S gene when receiving the transgene. Progeny of a female receives a modified transgene, which is never expressed.

H^{+} mice. In E36 H^{m} mice, the transgene was totally resistant to DNase I, showing that the chromatin region was in a closed configuration (Fig. 3). The fact that only one strain of mice shows a parental imprinting of the transgene suggests that mouse flanking sequences are responsible for this modification. Such a region may contain genes imprinted during gametogenesis and involved in the parental effect.[2,3] This prompted us to analyze in more detail the insertion locus. From a E36 DNA library, we cloned back the transgene and 2 kb of flanking mouse DNA. We isolated a unique probe that we named 4–12, located on chromosome 13 and hybridizing to a 5-kb *Msp*I/*Hpa*II

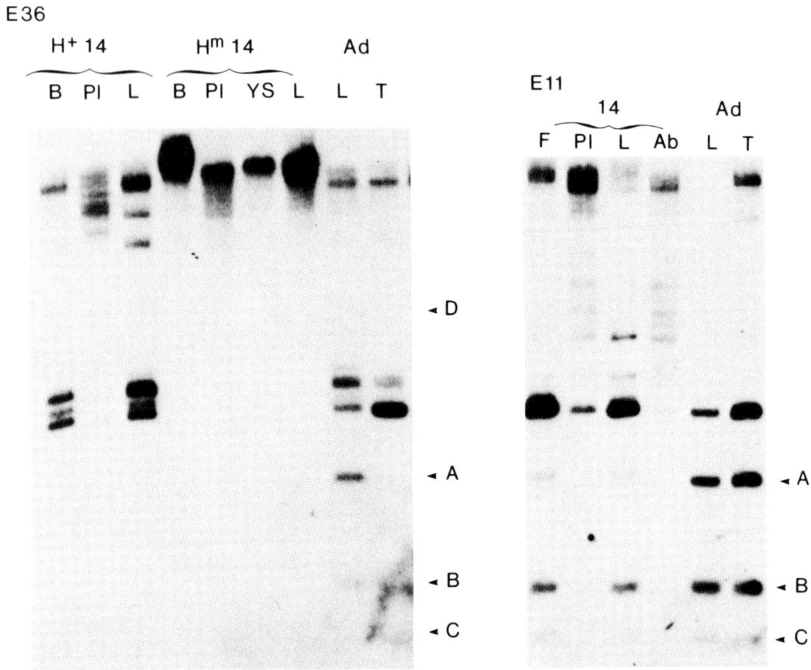

Fig. 2. Methylation status of HBV DNA from different tissues of E36 and E11 mice. E36 H^+ is an offspring of a male, and H^m is an offspring of a female. DNA was extracted from tissues of either 14-day-old fetuses or adult mice (Ad), digested with *Hpa*II and analyzed by Southern blotting and hybridization to a HBV probe: B (body), Pl (placenta), L (liver), YS (yolk sac), T (testis), Ab (abortive fetus). Bands A, B, and C are HBV fragments produced by *Hpa*II/*Msp*I digestion of the pAC2 plasmid injected in the transgenic mice. These bands are detected only in the liver of adult E36 mice and in the liver and testis of adult E11 mice. The DNA of H^m E36 mice is resistant to *Hpa*II digestion with a lesser degree in placenta.

fragment.[4] Using this probe, we analyzed the methylation status of mouse DNA extracted from different organs at different times of development. We found that the region is highly methylated in every adult tissue, and that in embryos, the degree of methylation is lowest in the placenta (Fig. 2). The restriction pattern allows differentiation between the male and female allele in a heterozygote transgenic mouse; the results demonstrate that the female allele is always hypermethylated, whether or not it contains the transgene.

We then screened a normal mouse DNA library with the 4–12 probe and isolated two clones spanning about 25 kb around the insertion point. A search for unique sequences cross-hybridizing with other species allowed the detection of a transcribed region on both sides of the transgene. This corresponds to two species of RNAs present in high amounts in the testis and

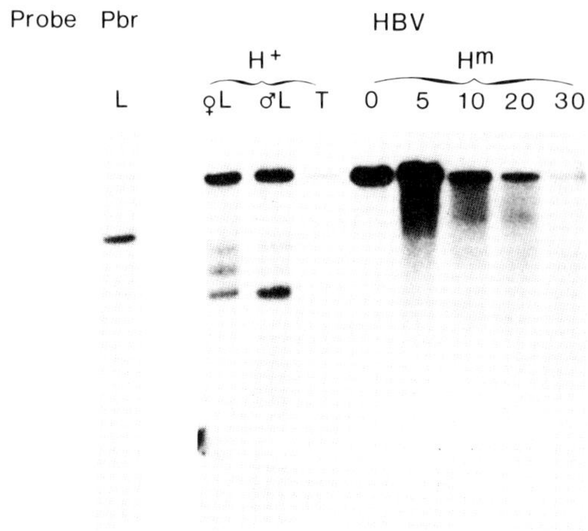

Fig. 3. DNase I sensitivity of HBV DNA in tissues of H^+ and H^m mice. Nuclei prepared from liver (L) or testis (T) samples were digested with limited amount of DNase I for 0 to 30 min. The DNA was then extracted and digested with *Pvu*II, which does not cut in HBV. The arrows on the left side show the DNA fragment produced by digestion at a DNase I site. Nuclei from a H^m mouse are resistant to DNase I digestion.

to three ubiquitous RNAs in smaller amounts. The testis-specific RNAs were not found in 10- and 18-day-old mice, suggesting that their synthesis may be restricted to postmeiotic germ cells. Indeed *in situ* hybridization locates the transcripts to the base of seminiferous tubules.

Our current hypothesis concerning the *de novo* methylation of the transgene and mouse adjacent regions is based on the presence of the testis-specific transcription unit next to the insertion. We think that such a gene may be inactivated in the female germ line, and that this may be followed by methylation. It is still not clear why hypomethylation of the locus is seen only in placenta.

REFERENCES

1. Farza, H., Salmon, A. M., Hadchouel, M., Moreau, J. L., Babinet, C., Tiollais, P., and Pourcel, C. (1987). *Proc. Natl. Acad. Sci. U.S.A.* **84,** 1187–1191.
2. Surani, M. A. H., and Barton, S. C. (1984). *Nature (London)* **303,** 548–550.
3. McGrath, J., and Solter, D. (1984). *Cell* **37,** 179–183.
4. Hadchouel, M., Farza, H., Simon, D., Tiollais, P., and Pourcel, C. (1987). *Nature (London)* **329,** 454–456.

8

A Role for Transcription in Antibody Switch Recombination

JOHANNA A. GRIFFIN[1] AND MARY E. REABAN[2]
Department of Microbiology
University of Alabama at Birmingham
Birmingham, Alabama 35294

INTRODUCTION

On the interaction of their surface-bound IgM with antigen, B cells are stimulated to differentiate and often to switch to the synthesis of a class of antibody other than the original IgM. The DNA recombination that coincides with this process occurs within sequences referred to as *switch regions* that range from 1 to 10 kilobases (kb) in length. The switch regions are located in the intron between the variable exon and the IgM constant exons, and 5′ of the constant exons encoding all other classes of antibody except IgD (Fig. 1) [for review, see (1)].

Results of experiments designed to elucidate the mechanisms of antibody class switching have not been consistent with the previous understanding of recombination processes. For instance, even though there is no significant interswitch region DNA sequence homology, switch recombination can occur in different switch regions on the two chromosomes of a single cell (2). Furthermore, the rearrangement does not occur in a defined DNA sequence, and the sites where the DNA rearrangement occurs differ from those that would be predicted to be involved in either classically defined homologous recombination or site-specific recombination. These data suggest that the recombination machinery is specific for some feature of the switch region other than DNA sequence.

1. *Present address:* Boehringer Ingelheim Pharmaceuticals, Inc., Ridgefield, Connecticut 06877

2. *Present address:* Molecular Genetics Program, Wadsworth Center for Laboratories and Research, New York State Department of Health, Albony, New York 12201

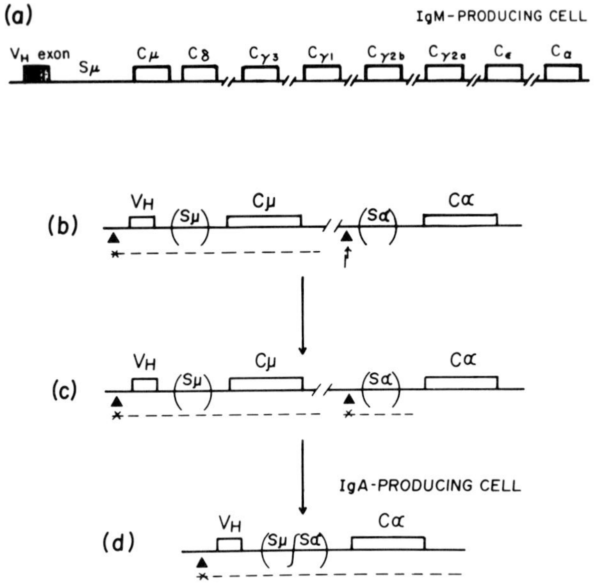

Fig. 1. Organization of the immunoglobulin heavy chain locus before and after switch recombination. The mouse heavy chain locus is composed of exons that encode eight different classes of antibody, Cμ, IgM; Cδ, IgD; Cγ3, IgG3; Cγ1, IgG1; Cγ2b, IgG2b; Cγ2a, IgG2a; Cε, IgE; and Cα, IgA (a). Since the active promoter of the variable (V_H) exon is initially located 5′ of Cμ, the first antibody expressed during B-lymphocyte differentiation is IgM. Before the switch to the synthesis of IgA, the promoter located 5′ of the Cα switch region (Sα) becomes transcriptionally active (b, c). Recombination then occurs between Sα and the Sμ sequences, resulting in the deletion of all of the DNA that originally separated the variable exon from the Cα exons (d).

Several tantalizing clues suggest that transcription may play a role in class switching. Before switching, transcriptional activation is specifically initiated from promoters located in or near the switch regions that are to be involved in the DNA recombination (3,4). Even though these promoters are conserved, they are deleted by switch recombination and never participate in the synthesis of antibody polypeptides. Recent studies predict that this transcription would promote supercoiling of the switch-region DNA (5,6). Since switch regions contain an abundance of sequences that have the potential to form alternative secondary structures when supercoiled (7), the transcriptional activation might induce the formation of alternative DNA structures in the recombination sites. Since these alternative structures would contain single-stranded stretches that would be complementary to their RNA transcripts, the RNA could anneal with those sites and stabilize the alternative structures. Earlier studies revealed that an RNA transcript could remain associated with its supercoiled DNA template (8). Finally, other studies have demonstrated that a single-stranded nucleic acid molecule has the potential to be recombinogenic (9) and that transcription can increase the rate of recombination (10).

Recently, we have tested some of our predictions experimentally. We have found that supercoiling of the IgA switch region (Sα) results in the formation of an intramolecular DNA triplex within Sα (11). Transcription of supercoiled Sα-containing plasmids enhances the formation of supercoil-relaxing structures in that switch region, and the RNA transcripts anneal with their templates and stabilize the topoisomers that result from the formation of alternative DNA structures (11). We think that these data provide additional support for the model that we have proposed for the mechanism of switch recombination.

RESULTS AND DISCUSSION

Effects of Transcription on Supercoiled Plasmids Containing IgA Switch-Region Sequences

We had previously found that a polypurine–polypyrimidine DNA repeat sequence of $(AGGAG)_{28}$ in the switch region 5′ of the IgA constant region formed an intramolecular triplex under negatively supercoiled conditions (Fig. 2) (12). To test the possibility that transcription might enhance the

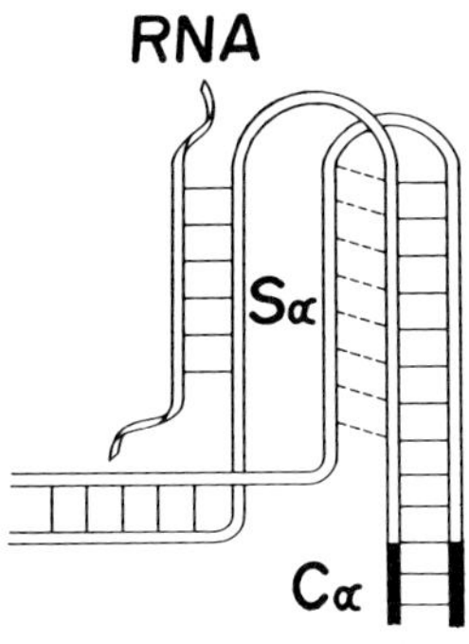

Fig. 2. Alternative DNA structure in negatively supercoiled IgA switch region. Sα contains a polypurine–polypyrimidine repeat $(AGGAG)_{28}$ that we have found capable of forming a supercoil-dependent triplex (11). A loop forms near the middle of the repeat, allowing the 3′ half of the pyrimidine strand to form Hoogsteen base pairs with the 3′ half of the purine strand. The 3′ half of the purine strand also remains Watson–Crick base-paired to the 5′ half of the pyrimidine strand. These interactions result in the 5′ half of the purine strand being unpaired and available to form the putative hybrid with the RNA transcripts.

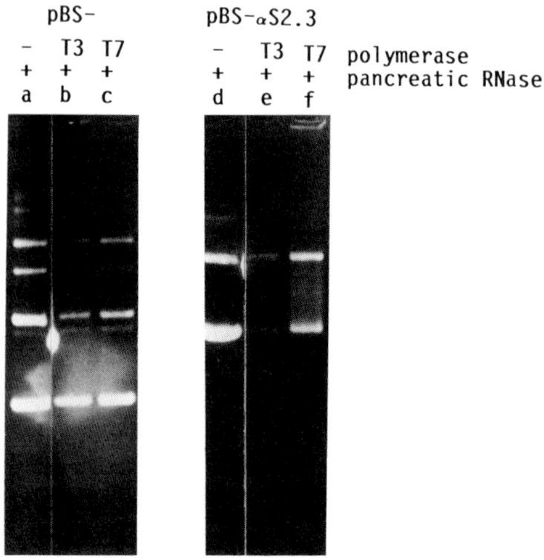

Fig. 3. Effects of transcription on supercoiled templates. Uncut plasmids were transcribed for 20 min, then phenol:$CHCl_3$ extracted and precipitated with ethanol. The samples were resuspended and treated with boiled pancreatic RNase, followed by another extraction and precipitation. Electrophoresis was performed on a TAE-buffered 1% agarose gel, and the DNA was visualized with ethidium bromide. There were no sites in the Sα insert that cause detectable stops or pauses of the T3 or T7 RNA polymerase (data not shown), nor were the DNA templates nicked or broken in the course of transcription. After transcription from the polypurine strand (T7) of pBS-αS2.3, a ladder of plasmid topoisomers appeared with superhelical densities ranging from that normally found in plasmids isolated from *Escherichia coli* to the completely relaxed form. The topoisomers were not seen following transcription from the polypyrimidine strand (T3) of pBS-αS2.3 or from either strand of the plasmid lacking the Sα insert.

formation of the supercoil-relaxing DNA triplex, we cloned Sα into a plasmid site flanked by the T3 and T7 bacteriophage RNA polymerase promoters. When RNA was transcribed from the polypurine strand of supercoiled plasmids, a ladder of progressively less negatively supercoiled topoisomers of the plasmid was generated (Fig. 3). Formation of the topoisomers was not seen following transcription from the polypyrimidine strand of Sα, from either strand of the plasmid lacking the Sα insert, or from linearized plasmids (data not shown). Therefore, transcription from a specific DNA strand appears to favor the formation of stable, supercoil-relaxing structures, such as the intramolecular triplex.

Role of RNA Transcripts in Stabilization of Topoisomers

We believe that the RNA transcript remains associated with its DNA template by base-pairing with the single-stranded region in the triplex, thus

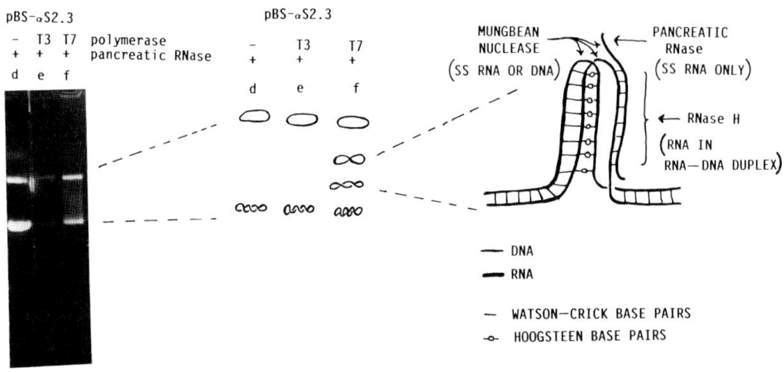

Fig. 4. Possible cause for the formation of the ladder of topoisomers. The topoisomers could be explained by the polypyrimidine RNA transcript remaining associated with its DNA template by base-pairing to the single-stranded polypurine strand generated by the formation of the DNA triplex, schematically diagrammed. Since the formation of the triplex removes negative supercoils, the binding of the polypyrimidine RNA would trap the plasmid in a less supercoiled form by stabilizing the structure. The appearance of a ladder of topoisomers could be owing to variations in length of the regions participating in the RNA–DNA duplex. Such a structure should be sensitive to pancreatic nuclease (single-stranded RNA), mung bean nuclease (single-stranded DNA or RNA), and RNase H (RNA in a RNA–DNA duplex), as indicated.

stabilizing that alternative structure. Since the triplex is composed of the 3′ portion of the polypurine strand simultaneously Watson–Crick base-paired with the 5′ portion of the polypyrimidine strand and Hoogsteen base-paired with the 3′ portion of the polypyrimidine strand, the 5′ portion of the polypurine strand is left unpaired. Therefore, only the polypyrimidine RNA transcript would be able to base-pair with the single-stranded region generated by triplex formation (Fig. 4). This analysis is supported by the strand-specific, transcription-dependent relaxation of supercoils described.

If our interpretation of these data were correct, treatment of the transcription reaction with RNase H would be expected to cause the molecules to revert to their original superhelical density owing to the selective degradation of the RNA of the enzyme in the RNA–DNA hybrids, whereas treatment with mung bean nuclease would relax the plasmids completely by digesting the single-stranded DNA exposed at the bend in the center of the triplex. Figure 5 reveals precisely these predicted results. On treatment with RNase H, the plasmids that had previously formed the topoisomeric ladder of gel bands migrated with the supercoiled form of the plasmid, indicating that an RNA–DNA hybrid had stabilized the topoisomers of intermediate superhelical density. The reversion of plasmids to the original supercoil density after RNase H treatment further indicated that the formation

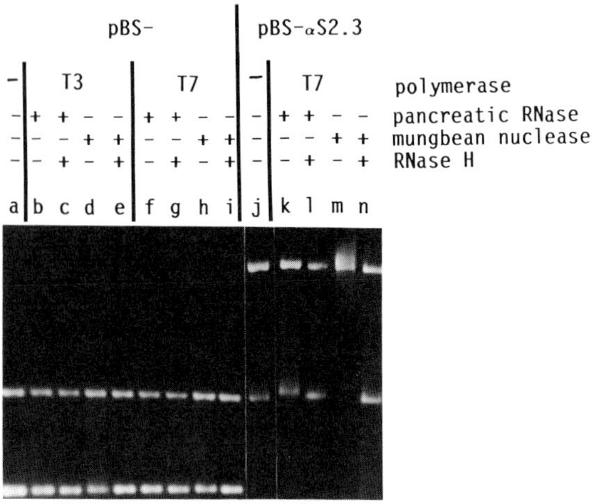

Fig. 5. Nuclease treatment of transcription products. Uncut plasmids were transcribed for 20 min, then phenol:$CHCl_3$ extracted. The samples were divided into aliquots and treated with either mung bean nuclease or boiled pancreatic RNase in the presence or absence of RNase H. Following extraction and precipitation, the samples were resuspended and electrophoresed in a TAE-buffered 1% agarose gel and visualized with ethidium bromide. On treatment with RNase H, the plasmids that had previously formed the topoisomeric ladder of gel bands migrated with the supercoiled form of the plasmid, and on treatment with mung bean nuclease, they migrated with the relaxed circular form of the plasmid. These nucleases had no effect on plasmids that did not form topoisomers following transcription (see Fig. 4 for nuclease specificities.)

of the topoisomers was the result of a conformational change rather than a change in the linking number. Additionally, all topoisomers treated with mung bean nuclease migrated with the relaxed circular form of the plasmid, indicating the presence of a region of single-stranded DNA in plasmids of all degrees of supercoiling.

Although the transcription-induced relaxation could conceivably be caused by simple R-loop formation, we favor the alternative interpretation described previously. If R-loops were forming without triplex formation, we would expect them to form independent of which strand was transcribed, and independent of the presence of the switch-region insert in the plasmid. We observed stable, relaxed topoisomers only after transcription from the polypurine sequence. Therefore, even if R-loop formation contributed to the formation of the topoisomers, the initial formation of a structure containing a single-stranded region appears to be required.

Although the polypyrimidine RNA has the potential to form the third strand of a DNA–DNA–RNA triplex, there are several arguments against the formation of such a complex in these experiments. The presence of such a structure would not cause the relaxation of supercoils that result in the ladder of topoisomers (Fig. 3), and the shifts in plasmid mobility are much too great to be caused by an increase in molecular weight contributed by an annealed strand of RNA. The formation of a DNA–DNA–RNA triplex, as well as any phenomena due to its formation, should occur following transcription of linear as well as of supercoiled plasmids. We observed no gel mobility shifts when linear plasmids were transcribed (data not shown). Finally, the RNA would be joined to the DNA by Hoogsteen, rather than by Watson–Crick base pairs and would not be susceptible to RNase H digestion, but the ladder of topoisomers completely disappeared on RNase H treatment. Therefore, it is unlikely that RNA forms the third strand of a triplex in these reactions.

MODEL FOR THE MECHANISM OF SWITCH RECOMBINATION

When a specific switch-region promoter becomes activated by a mediator in the microenvironment of the B lymphocyte, the transcriptional activity has multiple potential effects (Fig. 6). The act of transcription of the constrained loop of DNA would result in local transient supercoiling. To par-

TRANSCRIPTIONAL ACTIVATION OF SWITCH REGION

↓

TRANSIENT LOCALIZED SUPERCOILING
FORMATION OF ALTERNATIVE DNA STRUCTURES THAT RELAX SUPERCOILS

↓

ANNEALING OF RNA TRANSCRIPTS TO SINGLE-STRANDED SITES IN DNA TEMPLATE

↓

↓

INTERACTION OF PROTEIN OR RNP WITH STRUCTURE CONTAINING DNA-RNA HYBRID

↓

↓

↓

PROTEINS OR RNPS CUT AND RELIGATE DNA IN SWITCH REGIONS

Fig. 6. Model for switch recombination.

tially relax the negative supercoils, alternative DNA structures could then form in the transcriptionally active switch region. The RNA transcripts could base-pair with the single-stranded regions in the DNA that result from the formation of these structures. This formation of RNA–DNA hybrids would stabilize the alternative structure, thereby increasing the half-life of the structure, and thus its chances for participating in recombination.

Since both switch regions that participate in the recombination would be transcriptionally active, we would predict that both target sites would have RNA–DNA duplexes. The type of alternative structure that formed in a given switch region would not matter as long as it contained a single-stranded region complementary to the RNA transcript. It is possible that the RNA–DNA hybrid forms the unique substrate that will be recognized by the "recombinase." Recognition and interaction between the recombinases binding the two switch-region substrates would bring into proximity distant DNA sequences so that the cutting and religation that are the final steps in switch recombination can occur. It is also possible that the RNA becomes associated with B lymphocyte-specific proteins to form a ribonucleoprotein (RNP) particle that has the capability of cleaving and religating DNA. The RNA incorporated into this RNP could determine the substrate specificity of the particle in a manner analogous to that of other RNPs previously described (13).

Thus, our model predicts that class switching is directed by the transcriptional activation of specific promoters that induce the formation of alternative DNA structures stabilized by RNA, and that these RNA–DNA structures serve as the recognition substrates for the switch-recombination apparatus.

Note added in proof: Data, generated since this material was presented at the P & S Biomedical Symposium, have lead us to modify the interpretations put forth in this chapter. Please refer to *Nature (London)* **351**, 447–448 (1991), and subsequent publications.

ACKNOWLEDGMENTS

This work was supported by National Institutes of Health Research Grant (PO5-AI-23694; J. A. G.) and National Cancer Institute Training Grant (5-T32-CA09467) (M.E.R). J. A. G. is an Established Investigator of the American Heart Association (870111).

REFERENCES

1. Lutzker, S. G., and Alt, F. W. (1989). *In* "Mobile DNA" (M. M. Howe and D. E. Berg, eds.) Ch 31, American Society for Microbiology, Washington, D.C.
2. Webb, C. F., Cooper, M. D., Burrows, P. D., and Griffin, J. A. (1985). *Proc. Natl. Acad. Sci. U.S.A.* **82,** 5495–5499.

3. Stavnezer, J., Radcliffe, G., Lin, Y.-C., Nietupski, J., Berggren, L., Sitia, R., and Severinson, E. (1988). *Proc. Natl. Acad. Sci. U.S.A.* **85,** 7704–7708.
4. Lutzker, S., Rothman, P., Pollock, R., Coffman, R., and Alt, F. W. (1988). *Cell* **53,** 177–184.
5. Giaever, G. N., and Wang, J. C. (1988). **55,** 849–856.
6. Tsao, Y.-P., Wu, H.-Y., and Lui, L. F. (1989). *Cell* **56,** 111–118.
7. Shimizu, A., and Honjo, T. (1984). *Cell* **36,** 801–803.
8. Champoux, J. J., and McConaughy, B. L. (1975). *Biochemistry* **14,** 307–316.
9. Hotta, Y., Tabata, S., Stubbs, L., and Stern, H. (1985). *Cell* **40,** 785–793.
10. Thomas, B. J., and Rothstein, R. (1989). *Cell* **56,** 619–630.
11. Reaban, M. E., and Griffin, J. A. (1990). *Nature (London)* **348**, 342–344.
12. Collier, D. A., Griffin, J. A., and Wells, R. D. (1988). *J. Biol. Chem.* **263,** 7397–7405.
13. Mowry, K. L., and Steitz, J. A. (1988). *TIBS* **13,** 447–451.

PART III

GENETIC RECOMBINATION IN *DROSOPHILA* AND *CAENORHABDITIS*

9
Genetic and Molecular Studies of a Simple Meiotic System

R. SCOTT HAWLEY[1] AND ANNE E. ZITRON
Department of Molecular Genetics
Albert Einstein College of Medicine
Bronx, New York 10461

INTRODUCTION

In most eukaryotes, segregation of homologous chromosomes at the first meiotic division is mediated by homologous pairing and exchange (reviewed by Hawley, 1988). Chiasmata, the physical consequence of exchange, then serve to conjoin the two homologous centromeres, and thus assure their coorientation at metaphase I (Nicklas, 1974). However, some organisms also possess secondary segregational systems that ensure the disjunction of those chromosomes that, for one reason or another, do not recombine. One such system, known as distributive pairing, was discovered in *Drosophila melanogaster* females by Grell (1962a, 1976) and has subsequently been characterized at both the genetic and molecular levels.

Although exchange itself is sufficient to ensure disjunction in *D. melanogaster* females, the distributive system was initially postulated by Grell (1962a, 1962b) to explain the regular segregation of achiasmate chromosomes in *Drosophila*. As pointed out by Grell (1976), the existence of a distributive system allows us to understand the regular segregation of homologous chromosomes that have not undergone exchange. For example, the tiny 4th chromosomes pair during zygotene [as evidenced by the presence of 4th chromosomal synaptonemal complex (Carpenter, 1975)], and segregate normally, even though they never undergo meiotic exchange.

1. *Present address:* Department of Genetics, University of California at Davis, Davis, California 95616

The other larger chromosomes are usually involved in homologous exchanges that ensure their disjunction. However, normal segregation of these chromosomes occurs even in the rare cases in which exchange does not. For example, in females with structurally normal X chromosomes, the frequency of nonexchange X bivalents is almost 6%, whereas the frequency of spontaneous X nondisjunction is less than 0.5% (Bridges, 1916; Weinstein, 1936). Moreover, the frequency of X chromosomal nondisjunction is not greatly elevated in females that are heterozygous for inversions, even though such heterozygosity greatly decreases the frequency of X chromosomal exchange.

The cases listed refer to the segregation of homologous chromosomes that have failed to recombine. However, the literature is also replete with examples of the segregation of nonhomologous chromosomes that do not undergo exchange (Zimmering, 1976; Hawley, 1988). These segregational events are also mediated by the distributive system and can be highly efficient. For example an *attached*-X and a Y chromosome segregate from each other greater than 99.9% of the time without the benefit of crossing-over.

In the last few years, distributive segregation has become the focus of interest in our laboratory. Our approach has been to mutationally identify the genes that are required for this process, to clone those genes, and to determine their protein products. Because distributive segregation is, at least under some circumstances, a dispensable meiotic process, it has been possible to recover a number of mutations that are specifically defective in various aspects of this process. These mutations can be nominally divided into two classes: those that affect the choice of segregational partners but do not affect segregation *per se*, and those that allow normal partner choice while impairing segregation. In the last year we have cloned and analyzed, at the molecular level, one of the genes required for proper chromosome segregation (Zhang *et al.*, 1990).

We have used this combined genetic and molecular approach to focus on two basic questions. First, what is the mechanism by which segregational partners are chosen?; and second, what function(s) take the place of chiasmata to ensure the proper disjunction of these segregational partners? We shall consider these two questions in detail.

PARTNER CHOICE AND THE NATURE OF DISTRIBUTIVE PAIRING

It was Bridges (1916) who first suggested that the choice of partners for nonexchange segregations might follow a different set of rules than did choice of partners for exchange. When Bridges examined sex chromosome disjunction in XXY *Drosophila* females, he observed frequencies of X chromosomal nondisjunction that were much higher than those observed in corresponding XX controls. Bridges showed that these nondisjunctional events involved only nonexchange X chromosomes and resulted from XX ↔ Y segregational events. Bridges termed this process *secondary nondisjunction*. Because Y chromosomes exert only a very weak effect on X chromosomal exchange (see the following), Bridges concluded that the process that mediated pairing for exchange was different from the events that aligned nonexchange chromosomes for segregation.

Grell (1964, 1976) extended Bridges' argument by claiming that homology plays no role in establishing the patterns of distributive segregation. Rather she proposed that partner choice was determined by three factors, namely availability, chromosome size, and chromosome shape. Because of the crucial role this hypothesis has played in subsequent studies of distributive disjunction, it will be critically reviewed here. We shall first consider the seminal studies of Grell and then consider the later work by ourselves and others.

Grell's Early Studies

The first of Grell's so-called *rules for distributive pairing* was the rule of availability. The term *availability* refers to the observation that when only two nonexchange chromosomes are present in the cell, they will segregate from each other regardless of their identity. This is true even if the two chromosomes differ greatly in terms of size or shape (i.e., centromere position). Two decades of subsequent observations by a number of investigators have reconfirmed the validity of this principle. For example, *attached*-X and an *attached-4* chromosomes, both of which have lifetime passes to the distributive pairing pool, will virtually always segregate from each other when present in the same female.

Nonetheless, the argument that disjunction based on availability precludes a role of homology implies a rather naive view of chromosome structure. We now know that at least some of the major satellite sequences are present on more than one chromosome (Lohe and Roberts, 1988). Moreover, Miklos *et al.* (1988) have shown that the proximal euchromatin/β-heterochromatin of the X and *4*th chromosomes are highly homologous in terms of the number, composition, and distribution of certain classes of repeated sequences. Indeed, several workers have argued that the *4*th chromosome in *Drosophila melanogaster* is derived from the proximal X, and that the centromeric regions of the two chromosomes may share extensive homology (for review, see Hochman, 1976). Thus, to say that two chromosomes are nonhomologs does not necessarily imply that they lack homologous sequences in their heterochromatic regions. This is a point to which we shall return repeatedly.

Grell further argued that when three or more nonexchange chromosomes were available, the choice of partners is made on the basis of physical characteristics, such as size or shape, and not on the basis of homology.

Grell's argument that chromosomal size plays a role in the choice of disjunctional partners is based on the meiotic behavior of small X chromosomal fragments (free duplications). These chromosomes are usually quite small, relative to a normal X, and do not undergo exchange. Grell observed that in females bearing such a duplication, but which were otherwise genetically normal, some of these duplications induced high levels of *4*th chromosomal nondisjunction. As diagrammed, these nondisjunctions result from improper partner recognition, such that one *4*th chromosome pairs with and segregates from the duplication, whereas its homolog (the other *4*th chromosome) segregates at random.

Duplication ↔ *4*
←*4*→

Grell (1964) demonstrated elegantly that duplications that are similar in size (and also cytogenetic morphology, a point not noticed by Grell) to chromosome *4* induced very high levels of nondisjunction. In these instances, the level of *4*th chromosomal nondisjunction approached 33%, the value expected if pairings between a *4* and a duplication are as likely as *4–4* pairings. Grell also claimed that the ability of duplications that are larger or smaller than a normal *4*th chromosome to interfere with *4*th chromosome partner choice diminished as the difference in size increased (i.e., as a duplication became either larger or smaller, it induced progressively less *4*th chromosomal nondisjunction). Thus, according to Grell, any duplication can interfere with *4*th chromosomal disjunction, but the probability of such events is determined not by the DNA sequences it carries, but solely by its chromosomal size.

Our own preliminary observations on this point paint a very different picture. Although we also observe high levels of nondisjunction in the case of those duplications that are similar in size and morphology to chromosome *4*, little or no effect on *4*th chromosomal segregation is observed with the larger or smaller duplications [Hawley, Zitron, and New, unpublished observations (1991)]. Given that our observations are corroborated by those of O'Tousa (1982), the discrepancy with those of Grell is unexplained.

We suspect that those duplications that do induce very high frequencies of *4*th chromosomal nondisjunction, and that are similar in size and morphology to the *4*th chromosome, are not free duplications at all. Given the manner in which these chromosomes were constructed, it is entirely possible that these small chromosomes are the result of translocations in which the euchromatic tip of the X has been captured by the centromeric heterochromatin of chromosome *4*. This hypothesis is currently being tested using both genetic and molecular methodologies.

If these so-called *duplications* are, in fact, half-translocations between the X and *4*th chromosomes, then the observed segregational events argue for a role of heterochromatic homology in distributive disjunction, rather than for a role of size. Thus, given the existence of at least one alternative explanation for Grell's data, the so-called *size rule* can be considered to be only a suggestion.

Grell's argument that chromosomal shape plays a role in determining distributive disjunctions is based on the observation that when two acrocentric and one metacentric chromosome (for example, two X chromosomes and a Y) are nonexchange, the metacentric chromosome will disjoin efficiently from the two acrocentric chromosomes (for review, see Grell, 1976). Based on a model proposed by Cooper (1948), it is believed that such segregations result from the formation of a trivalent in which each arm of the metacentric is associated with one of the acrocentric chromosomes. However, the formation of trivalents, especially those involving chromosomes like the X and the Y (which share heterochromatic homology), does not speak for or against a role for homology in determining distributive associations.

In fact, Grell (1962b) herself presented strong evidence for a role of homology in the control of secondary nondisjunction (i.e., XX ↔ Y segregation). Grell (1962b) noted that when two isosequential X chromosomes were nonexchange, they were much less likely to undergo secondary nondisjunction than were two X chromosomes that differed by a structural aberration (such as an inversion). This observation, along with the failure of a free Y chromosome to disjoin from the two acrocentric *4*th chromosomes, led even Grell to conclude that "homology plays a major role in distributive segregation" (Grell, 1962b).

The Work of Others

In addition to Grell's studies, those of several other workers have also suggested that the parameters of size and shape are insufficient to fully explain partner choice, and further supported a role of homology in distributive disjunctions. For a detailed review of the earlier work on this subject, see Novitski and Puro (1978).

Perhaps the most relevant of these efforts are the works of Gershenson (1940) and of Lindsley and Sandler (1958). These investigators examined X chromosome segregation in females that carried a small X chromosomal free duplication, and that were also heterozygous for an exchange-suppressing X chromosome inversion. (Note: in the discussion that follows, we shall be considering free duplications whose genetic makeup or morphology indicates that they are indeed X chromosomal free duplications.)

In both studies these duplications were shown to be capable of inducing significant frequencies of X nondisjunction as a consequence of XX ↔ *Duplication* segregational events. Since the two X chromosomes are much more similar in size to each other than either is to the duplication, the observed nondisjunctional events cannot be explained by the size rule. Given that all three chromosomes are acrocentrics, the shape rule is also of little help in explaining the data. These workers argued, we think correctly, that the observed segregational events were the result of homology-dependent associations between the X chromosomal free duplication and the full-length X chromosomes from which it was derived.

Our own data strongly corroborate this hypothesis. In an experiment similar to that of Lindsley and Sandler (1958), but in which both X and *4*th chromosomal disjunction could be measured, we have shown that a free X duplication (approximately twice the size of chromosome *4*) induces more than twice as many XX ↔ *Duplication* disjunctions as it does *44* ↔ *Duplication* disjunctions (Zitron and Hawley, 1989). Again, given that this duplication, which is known to be derived from the X, is much closer in size to chromosome *4* than to the X, these data are not consistent with the size rule. Slightly larger X chromosomal duplications induce very high levels of X chromosomal nondisjunction with little or no effect on chromosome *4* [Lindsley and Sandler, 1958; our unpublished data (1991)]. These observations suggest that the X and the duplication chromosomes recognize each other by virtue of their relatedness, even in an instance in which the homology present is insufficient to allow exchange.

Moreover, we have recently analyzed a new meiotic mutation, *Axs*, which is specifically defective in the recognition of homology within the distributive system (Zitron and Hawley, 1989). Although *Axs* has no effect either on the frequency of exchange or on exchange-mediated disjunction, it does increase the frequency with which nonexchange chromosomes nondisjoin. These nondisjunctional events are not the consequence of segregational failures, but rather reflect a high frequency of improper, or nonhomologous segregational events. For example, the X chromosomes nondisjoin at frequencies of nearly 40% in *Axs*/*Axs* females in which exchange is suppressed, and a large number of these events are owing to X ↔ *4* segregations. A mutation with a similar phenotype (*ald*) has been characterized by O'Tousa (1982).

For both *Axs* and *ald* the basic phenotypic defect may be summarized as follows: homologous distributive segregations are decreased in frequency and replaced by nonhomologous segregations. Based on these data, we propose that the wild-type *Axs* and *ald* gene products act to promote homologous associations and to diminish the probability of nonhomologous pairings. It is difficult to understand the existence of such functions or the observed mutant phenotypes if the distributive system is truly homology independent.

We argue, therefore, that homolog recognition plays a major role in the choice of partners within the distributive system. We further propose that when only two nonhomologs are available, disjunctions are mediated by homologous sequences (quite possibly heterochromatic satellite DNA sequences) that are shared between the two chromosomes. The concept of homology-dependent distributive associations allows us, following Novitski (1964, 1975, 1978), to envision the following rather speculative model of meiosis in *D. melanogaster* females.

A Revised Model of Meiosis

In early prophase, homologous chromosome pairing takes place between both euchromatic and heterochromatic regions. The euchromatic pairings provide an opportunity for exchange between homologs. We propose that the heterochromatic pairing events may take place between both homologous and nonhomologous chromosomes, and that these chromosomal alignments or associations will, in the absence of chiasmata, determine distributive segregational events. Since distributively segregating chromosomes move to the poles much earlier than do chiasmate bivalents (Puro and Nokkala, 1977; but see the following), these associations will be voided by exchange.

If this model is correct, then the term *distributive pairing* corresponds in reality to heterochromatic pairings involving both homologous and nonhomologous chromosomes. We propose that the cell attempts to minimize the number of unpaired or univalent chromosomes, which are highly unstable and likely to be lost (for review, see Hawley, 1988). We also propose that the cell attempts to maximize the fidelity of these associations by pairing those chromosomal regions that share the most extensive homology. These two processes help to ensure that each chromosomal arm will be paired.

The question then arises as to whether or not the proposed heterochromatic pairings that facilitate distributive segregation are entirely independent of the euchromatic pairings that facilitate exchange. Several lines of evidence strongly suggest that the two systems are not independent. First, two separate studies strongly suggest that XX-Y or XX-*44* associations can result in region-specific exchange reductions observed only at the base of the X chromosomes, as if X-*4* or X-Y pairings can compete with or impede X-X pairing for exchange (Merriam, 1964; Zitron and Hawley, 1989; but see

Hawley, 1989). Second, several workers have also shown that exchange chromosomes can, under some rather special circumstances, participate in distributive segregation events. For example, Carpenter has shown that exchange X chromosomes occasionally undergo secondary nondisjunction in XXY females (Carpenter, 1973). Similarly, Novitski (1978) has shown that the Y chromosomal arms of a Y-autosomal translocation can still orient the segregation of two nonexchange X chromosomes, even when the autosomal material connected to the Y is involved in exchange. Finally, the *mei-S51* mutation both disrupts the exchange process and interferes with partner choice for distributive disjunction (Robbins, 1971), suggesting that the two processes share at least one common component.

THE SEGREGATION OF NONEXCHANGE CHROMOSOMES

Even if the basis of distributive pairing can be shown to lie in a conventional process, such as homology recognition, the question still remains: what functions substitute for chiasmata, and thus serve to coorient the centromeres of nonexchange chromosomes? One clue to the mechanism(s) of distributive segregation comes from cytological observations of prometaphase in *Drosophila* females (Puro and Nokkala, 1977). Nonexchange chromosomes are first observed to be physically conjoined and then to coorient on the same arc of the spindle. These chromosomes then undergo precocious poleward migration during the early stages of spindle formation. Thus, by the time a mature spindle is completed at metaphase, the achiasmate chromosomes are at a position intermediate between the plate and the poles. This behavior is in sharp contrast to that of exchange bivalents, which are maintained by chiasmata at the metaphase plate.

Two genes are known to produce proteins that are required for normal distributive segregation. Mutations at both these loci, known as *nod* and *ncd*, induce high levels of nondisjunction involving nonexchange chromosomes. Because the *nod* locus is specifically required for distributive segregation, and because this locus has been the focus of our laboratory's interests, we shall consider this locus first and then return to a discussion of *ncd*.

The *nod* Gene

The *nod*, or *no distributive disjunction*, mutation was recovered by A. T. C. Carpenter and B. S. Baker and reported in Carpenter (1973). Mutations at the *nod* locus do not affect either exchange or the behavior of exchange chromosomes; rather, they specifically impair the segregation of achiasmate chromosomes (Carpenter, 1973; Zhang and Hawley, 1990).

The most pronounced effect of *nod* on female meiosis is on the always achiasmate *4*th chromosomes. Fourth chromosome nondisjunction frequencies approach 90% in *nod*/*nod* females, with the vast majority of exceptional gametes having lost the *4*th chromosome. Mutations at the *nod* locus also induce random X chromosomal disjunction in those meiocytes containing nonexchange X chromosomes (Carpenter, 1973; Zhang and Hawley, 1990). Both partner choice and centromere coorientation occur normally in *nod* females, suggesting that the defect is specific to segregation (Carpenter, 1973).

We have cloned the *nod* locus, examined its transcription patterns, and determined its coding sequence (Zhang *et al.*, 1990). The N-terminal domain of the predicted coding sequence has strong amino acid similarity to the mechanochemical domain of kinesin heavy chain and thus may be a microtubule motor. By analogy with kinesin (Vale *et al.*, 1985), we have proposed that the *nod* kinesinlike protein acts by generating plus-end directed (that is to say, toward the metaphase plate) forces that serve to balance the poleward forces exerted by centromeres. This hypothesis is supported by recent cytological studies that demonstrate that in females homozygous for the *nod* mutation, nonexchange chromosomes drift away from the developing spindle and either are lost or reassociate at random (Theurkauf and Hawley, 1992).

Until our immunolocalization studies are completed, we can only speculate as to the site of *nod* function. One possibility is that the *nod* protein is a component of the so-called *polar ejection force*, which apparently acts along the whole length of a chromosome to propel it away from the spindle pole (Rieder *et al.*, 1986; Salmon, 1989; but see Carpenter, 1991). Alternatively, given that Hyman and Mitchison (1991) have recently demonstrated that the centromere is capable of moving chromosomes both away from and toward the pole, one could imagine that the *nod* protein is a centromere component that either pushes away from the pole or even is simply a brake.

Although the *nod* protein is required only for meiosis, and therefore only for distributive disjunction, two lines of evidence suggest that this protein also plays a role in mitosis. First, as shown by Zhang *et al.* (1990), the *nod* transcript is present throughout development. Second, an antimorphic allele of *nod*, known as nod^{DTW}, affects both meiotic and mitotic chromosome behavior (Wright, 1974; Rasooly *et al.*, 1991). We have recently shown that the nod^{DTW} lesion results in an alteration in the putative adenosine triphosphate (ATP)-binding domain (Rasooly *et al.*, 1991). By analogy with a similar mutation in the yeast kinesinlike protein KAR3 (Meluh and Rose,

1990), we propose that this mutation results in a "rigor-binding" defect. The finding that the *nod* protein is present but not required in mitotically dividing cells suggests that, at least mitotically, the *nod* protein is either functionally or genetically redundant.

The *ncd* Gene

Mutations at the *ncd* locus also induce high frequencies of meiotic chromosome loss and nondisjunction (Davis, 1969; Sequeira *et al.*, 1989). Although *ncd* mutations induce both exchange and nonexchange chromosomes to undergo nondisjunction at high frequencies, nonexchange chromosomes undergo nondisjunction and loss much more frequently than do exchange chromosomes (Carpenter, 1973).

The physical basis of the *ncd* defect is the abnormal segregation of chromosomes by disorganized spindles with abnormal numbers of poles (Kimble and Church, 1983). Recently the *ncd* locus has been shown to encode a kinesinlike protein, which displays both minus-end (i.e., poleward) motor activity and the ability to bundle microtubules (Endow *et al.*, 1990; McDonald and Goldstein, 1990; Walker *et al.*, 1990; McDonald *et al.*, 1990).

Thus, the *nod* protein apparently acts to hold nonexchange chromosomes on the developing spindle, and thus in some sense compensates for a lack of chiasma function; whereas the *ncd* protein is essential for the development of the spindle (such that the lack of *ncd* function makes nonexchange chromosomes more sensitive to nondisjunction or loss). These two processes are sufficiently interrelated that recessive loss-of-function mutations at the *ncd* locus act as dominant enhancers of mutations at the *nod* locus (Knowles and Hawley, 1991).

SUMMARY

In the preceding pages we have detailed a molecular, cytological, and genetic analysis of a relatively simple meiotic process. We argue that a rather simple picture of this process is beginning to emerge, and that distributive associations are not magical unions based on size or shape, but rather reflect a search for homology, even when those homologous sequences are on other chromosomes. We have further shown that the functions of chiasmata are replaced in the distributive system by proteins such as the *nod* and *ncd* kinesinlike proteins. Whereas such ideas and findings are but a bare beginning, we think they reflect at least a first step toward a detailed molecular understanding of the meiotic process.

REFERENCES

Bridges, C. B. (1916). *Genetics* **1,** 1–51, 107–163.

Carpenter, A. T. C. (1973). *Genetics* **73,** 393–428.

Carpenter, A. T. C. (1975). *Chromosoma* **51,** 157–182.

Carpenter, A. T. C. (1991). *Cell* **64,** 1–6.

Cooper, K. W. (1948). *Proc. Natl. Acad. Sci. U.S.A.* **34,** 179–187.

Davis, D. G. (1969). *Genetics* **61,** 577–594.

Endow, S. A., Henikoff, S., and Soler-Niedzella, L. (1990). *Nature (London)* **345,** 81–83.

Gershenson, S. (1940). *Videnskap. Akad. Nauk. U.R.R.S.* **3,** 16.

Grell, R. F. (1962a). *Proc. Natl. Acad. Sci. U.S.A.* **48,** 165–172.

Grell, R. F. (1962b). *Genetics* **50,** 150–166.

Grell, R. F. (1964). *Proc. Natl. Acad. Sci. U.S.A.* **52,** 226–232.

Grell, R. F. (1976). *In* "The Genetics and Biology of *Drosophila*," (M. Ashburner and E. Novitski, eds.), Vol. 1A, pp. 435–486. Academic Press, New York.

Hawley, R. S. (1988). *In* "Genetic Recombination," (R. Kucherlapati, and G. R. Smith, eds.), pp. 497–525. ASM Press, Washington, D.C.

Hawley, R. S. (1989). *In* "Molecular and Cytogenetic Studies of Nondisjunction," (T. Hassold ed.), pp. 277–302. Alan R. Liss, New York.

Hochman, B. (1976). *In* "The Genetics and Biology of Drosophila," (M. Ashburner and E. Novitski, eds.) Vol. 1B, pp. 903–928. Academic Press, New York.

Hyman, A. A., and Mitchison, T. G. (1991). *Nature (London)* **351,** 206–211.

Kimble, M., and Church, K. (1983). *J. Cell Sci.* **62,** 302–318.

Knowles, B. A., and Hawley, R. S. (1991). *Proc. Natl. Acad. Sci. U.S.A.* In press.

Lindsley, D. L., and Sandler, L. (1958). *Genetics* **43,** 409–425.

Lohe, A. R., and Roberts, P. (1988). *In* "Heterochromatin: Molecular and Structural Aspects" (R. S. Verma, ed.), pp. 148–186. Cambridge University Press, Cambridge, England.

Meluh, P. B., and Rose, M. D. (1990). *Cell,* **60,** 1029–1041.

Merriam, J. R. (1967). *Genetics* **57,** 409–425.

Miklos, G. L. G., Yamamoto, M. T., Davies, J., and Pirrotta, V. (1988). *Proc. Natl. Acad. Sci. U.S.A.* **85,** 2051–2055.

McDonald, H. B., and Goldstein, L. S. B. (1990). *Cell* **61,** 991–1000.

McDonald, H. B., Russell, J. S., and Goldstein, L. S. B. (1990). *Cell* **63,** 1159–1165.

Novitski, E. (1964). *Genetics* **50,** 1449–1451.

Novitski, E. (1975). *Genetics* **79,** 63–71.

Novitski, E. (1978). *Genetics* **88,** 499–503.

Novitski, E., and Puro, J. (1978). *Hereditas* **89,** 51–67.

Nicklas, R. B. (1974). *Genetics* **78,** 205–213.

O'Tousa, J. (1982). *Genetics* **102,** 503–524.

Puro, J., and Nokkala, S. (1977). *Chromosoma* **21,** 1–16.

Rasooly, R., New, C., Zhang, P., Hawley, R. S., and Baker, B. S. (1991). *Genetics* **129,** 409–422.

Rieder, C. L., Davison, E. A., Jensen, L. C. W., Casimeris, L., and Salmon, E. D. (1986). *J. Cell Biol.* **103,** 581–591.

Robbins, L. G. (1971). *Mol. Gen. Genet.* **110,** 144–166.

Salmon, E. D. (1989). *In* "Mitosis" (J. S. Hymans and B. R. Brinkley, eds.), pp. 119–181. Academic Press, New York.

Sequeira, W., Nelson, C. R., and Szauter, P. (1989). *Genetics* **123,** 511–524.

Theurkauf, W., and Hawley, R. S. (1992). *Cell. Biol.* In press.

Walker, R. A., Salmon, E. D., and Endow, S. A. (1990). *Nature (London)* **347,** 780–782.
Weinstein, A. (1936). *Genetics* **21,** 155–199.
Wright, T. R. F. (1974). *Genetics* **76,** 511–536.
Vale, R. D., Reese, T. S., and Sheetz, M. P. (1985). *Cell* **42,** 39–50.
Zhang, P., and Hawley, R. S. (1990). *Genetics* **125,** 115–127.
Zhang, P., Knowles, B. A., Goldstein, L. S. B., and Hawley, R. S. (1990). *Cell* **62,** 1053–1062.
Zimmering, S. (1976). *In* "The Genetics and Biology of *Drosophila*" (M. Ashburner and E. Novitski, eds.), Vol. 1B, pp. 569–613. Academic Press, New York.
Zitron, A. E., and Hawley, R. S. (1989). *Genetics* **122,** 801–821.

10

Diversity among *Drosophila* Transposable Elements and in Their Effects on Gene Expression

STEPHEN M. MOUNT
Department of Biological Sciences
Columbia University
New York, New York 10027

INTRODUCTION

Transposable elements are widespread in the genomes of higher eukaryotes, where they are responsible for a variety of genetic phenomena, including insertions, deletions, and chromosomal rearrangements. These elements are structurally diverse, and can be divided into several groups. The single eukaryotic species in which eukaryotic transposable elements are best described, with regard to their diversity, the nature of the mutations caused by their insertion, and the conditions that lead to transposition, is the fruit fly *Drosophila melanogaster.* This chapter will briefly summarize the major classes of transposable elements found in the *Drosophila* genome (almost all of which have been observed in other multicellular eukaryotes), and will also review current research into the molecular mechanisms by which the activities of transposable elements alter gene expression. Eukaryotic transposable elements have been the subject of extensive research in recent years, and a chapter of this length can provide only a general overview of the subject. Many of the references in this article will be to reviews, and I have drawn heavily on more substantial earlier reviews, which the interested reader should consult directly (1–6).

CLASSES OF TRANSPOSABLE ELEMENTS

Transposable elements in *Drosophila* can be divided into several groups based on their structural features (Fig. 1; 5–8). The major division is between those elements, known as retrotransposons, that transpose through a mechanism involving an RNA intermediate, and those elements whose transposition does not involve an RNA intermediate. Retrotransposons include long tandem repeat (LTR) retrotransposons (also known as copialike, or retroviruslike, elements) and non-LTR retrotransposons. The DNA elements can be divided into those with short inverted repeats and those with long inverted repeats. Almost all repeated elements fall into one of these four classes, and subdivision within these classes is possible. Interestingly, closely related transposons have been found in the genomes of taxonomically diverse species, indicating that these transposons are either very old or are capable of transmission between distantly related species.

LTR Retrotransposons

The largest class of repeated sequences in the *Drosophila* genome is the LTR retrotransposon family (reviewed in 4 and 9). Elements in this family are sometimes known as copialike elements (7), based on the fact that copia was the first such element to be extensively characterized, but may be better described as retroviruslike, based on their considerable similarity to retroviruses. Like retroviral proviruses, elements in this group are characterized by long direct tandem repeats (LTRs; see Fig. 1); the sizes of 11 *Drosophila* LTR sequences cataloged by Fawcett and Finnegan (4) vary between 266 and 512 nucleotides. Other features also indicate that the mechanism of transposition of these *Drosophila* elements is similar in detail to the mechanism of retroviral replication (reviewed in 4,5,9, and 10). These include (1) the generation of three to six nucleotide target-site duplications on insertion; (2) complementarity to cellular transfer RNAs (tRNAs) immediately adjacent to the 5′ LTR and a short purine-rich stretch adjacent to the 3′ LTR; (3) terminally redundant RNAs generated via transcription from one LTR to the other; (4) open reading frames with sequence similarity to retroviral proteins having known roles in the replication process; (5) very short inverted repeats at the termini of the LTRs; and (6) intermediates (or by-products) such as circular DNA forms, RNA–DNA hybrids, and viruslike particles containing reverse transcriptase activity. These properties are shared not only with retroviruses but also with transposable elements from other species, including the yeast element Ty1, which has been shown to transpose via an RNA intermediate (11).

Also like retroviruses, most of these elements begin with the sequence TG

LTR retrotransposons	Long (200–500 n.) directly oriented terminal repeats
	Related to retroviruses and retrotransposons in many species
TG...CA:	Copia, 1731, 3S18/BEL, B104/roo, 412, mdg1, Stalker (71)
ANT...ANT:	Flea, 297, 17.6, HMS Beagle, mdg4/gypsy, springer, opus
non-LTR retrotransposons	No repeats. One end A-rich
	Related to mammalian LINEs, the protozoan element Ingi and the Cin4 element of *Zea mays*
	Jockey, I, F, G, D, wallaby (77), R1Dm, R2Dm
Short inverted repeat elements	Short inverted repeats
	Related to Tc1 in *Caenorhabditis elegans*
	P, hobo, pogo, HB, mariner
Long inverted repeat elements	Long inverted repeats
	Related to sea urchin TU elements
	FB, BS

Fig. 1. Each of the four major classes of transposable elements is briefly summarized, a simple drawing indicates structural features common to elements in that class, and examples are given. Related elements found in species distantly related to *Drosophila* are mentioned. This information is reviewed in 1–9. Additional references and a more complete description can be found in the text. n, Nucleotide.

(so that the LTR, and the entire element, can be represented by TG . . . CA). However, there is a distinct subgroup (5 of the 11 sequences in refs. 4 and 9) with the sequence ANT . . . ANT (see Fig. 1), and I refer to the elements in these groups as TG and ANT elements. Interestingly, all characterized insertions of ANT elements have been into the sequence TATATA, or into similar sequences (e.g., TATACA or TACATA). This stands in contrast to observations with copia and other TG elements, where no similarity has been observed among many insertion sites (9). Such a connection between LTR termini and target-site specificity is reasonable because both would be expected to reflect the activity of the integrase that joins DNA intermediates with the chromosomal insertion site. Further exploration of these properties should be made possible by studying the integration of an ANT element *in vitro*. Such *in vitro* integration reactions have now been successfully performed with three retroviruses (12–15) and the yeast element Ty (16).

Non-LTR Retrotransposons

The second major class of retrotransposons is that containing elements related to mammalian long interspersed element (LINEs) (17,18) and short interspersed element (SINEs), including processed pseudogenes (19). These elements are characterized by the absence of repeated termini, variable (but relatively long) target-site duplications, and the occurrence at one end of an A-rich sequence. They have been reviewed most recently by Finnegan (20,21) and by Boeke and Corces (6), who refer to them as *poly(A)-type retrotransposons*. Those that have been sequenced generally have two open reading frames on the strand bearing oligo(A), and show similarity at the protein sequence level to reverse transcriptases (8,23,24). Transposition is believed to proceed after reverse transcription of an RNA initiated at the 5′ end of the element, and such an RNA has been identified for jockey and I (25,26). Indeed, jockey has been shown to have an internal promoter (25), as would be necessary if the promoter were to transpose with the element. Most individual examples of non-LTR retrotransposons are truncated at their 5′ ends, a property they share with processed pseudogenes found in mammals. This truncation has been attributed to incomplete reverse transcription.

Elements with Short Inverted Repeats

Five transposable elements with short inverted repeats have been described in *Drosophila* (4,27,28). These are P (reviewed in 29), hobo (reviewed in 30), pogo (O'Hare, unpublished, described in 4), HB (27) and mariner (28). Of these, the best studied is the P element, which has been thoroughly and recently reviewed (29,31,32). The mariner element is found, not in *Drosophila melanogaster*, but in its sibling species *D. mauritiana* and *D. simulans*. Terminal short inverted repeats are 31 nucleotides in the case of P, 21–23 for pogo, 28 for mariner, and 12 for hobo. In addition, both P and hobo are similar in that both generate eight-bp target-site duplications on transposition, whereas mariner and pogo may both generate two-bp duplications of the target-site sequence TA. The HB appears to have inverted repeats of about 30 nucleotides. The exact length of the HB inverted repeat and the size of HB target-site duplications are uncertain because HB has never been seen to transpose. Many of these properties are shared by similar elements in other species, particularly plants (5,29,33,34).

Unlike retrotransposons, these elements appear to be able to bring about a variety of alterations at the DNA level. I shall describe the activities of P elements, but it is becoming clear that many of these properties are shared by hobo elements (30) and may be more general. Structurally, P elements consist of complete autonomous 3-kb elements, which carry all of the functions needed for transposition, and nonautonomous elements, which carry those sequences required in *cis* for transposition and can transpose only when in the presence of a source of transposase, such as a complete P element. In natural populations, most nonautonomous elements are related to complete elements by internal deletions. In the laboratory, nonautonomous elements generally carry marker genes, which have been placed there by investigators. Such marked elements are now frequently used in germ-line transformation of *Drosophila* (35,36). By providing or removing a source of transposase, P element transposition can be controlled genetically. A common choice for experimental purposes is P[ry^+, Δ2-3](99B), an unusually stable transposase-producing element described by Robertson *et al.* (37).

Transposition can be followed either by observing mutations caused by P element insertion or by noting the change of linkage of an element carrying a phenotypic marker. In general, transposition rates are high, ranging from 0.01 to 0.25 new insertions per element per generation (29). The P elements can also promote events other than forward transposition. These include recombination in males (which lack meiotic recombination in *Drosophila*), chromosomal rearrangements (which typically involve breakpoints at the site of P elements; 38) and the occurrence of deletions adjacent to a P element insertion. Particularly common are alterations in the sequence of the initial P element. These include excision, internal deletion, gene conversion (39), and the insertion of a second element within, or very near to, a preexisting element (29). Recently, Engels *et al.* (32) have shown that the frequency of precise excision is greatly stimulated by the presence of a homologous chromosome, and have interpreted this and other data as supporting a "cut-and-paste" model of P element transposition like that previously described for bacterial elements such as Tn*10* (31,40). Central to the model is the idea that transposase-dependent rearrangements can be explained as resulting from double-strand gap repair.

Elements with Long Inverted Repeats

Another class of *Drosophila* element distinct from the retrotransposon family is that consisting of elements with long inverted repeats (reviewed in 9). Although such elements are found in a number of species, *Drosophila* contains only one well-characterized family, the FB element family. The BS element is known from a single insertion mutation (41). Each occurrence of an FB element includes long inverted repeats flanking a variable central region. The inverted repeats themselves consist of variable numbers of tandemly, but imperfectly, repeated copies of a 31-bp sequence and, most distally, less repetitious sequence (42,43). The FB elements are associated with a number of extremely unstable alleles including *white-crimson* (44,45), *white-DZL* (46,47), and the Ising TE (48,49), a large composite transposon in which a chromosomal segment of several hundred kb is mobilized by flanking FB elements.

Relationships between Transposable Element Classes in *Drosophila* and in Other Species

Each of the four major classes of *Drosophila* transposable elements is also found in taxonomically diverse eukaryotic species (see Fig. 1). The long inverted repeat class in *Drosophila* is represented by TU elements in sea urchin species (50). Elements with short inverted repeats are extremely diverse, and include most bacterial transposable elements (1). In addition, HB is similar to Tc1, an element found in the nematode worm *Caenorhabditis elegans,* at two levels (51): there is 24% identity at the amino acid level in a region of 273 amino acids, and there are 16 identical nucleotides in the 24 terminal nucleotides of their inverted repeats.

Similar cross-taxa similarities are common among retrotransposons. There is sufficient conservation of critical amino acids among reverse transcriptases for these elements to be classified taxonomically according to the sequence of that protein (8,24). As expected from the categories described above, the major division among reverse transcriptases is between those encoded by elements with LTRs and those encoded by elements without LTRs. The non-LTR class is related to reverse transcriptases encoded by open reading frames associated with group II introns of fungi and msDNA in bacteria. The LTR class can be divided into LTR retrotransposons, hepadnaviruses, caulimoviruses, and retroviruses. As expected, there are many cases of closely related elements found within a single host species. For example, the copia and 1731 retrotransposons of *D. melanogaster* are similar, with roughly 30% amino acid identity (52). In addition, elements in one species of *Drosophila* are closely related to elements found in related species of *Drosophila.* For example, the sibling elements 17.6 and 297 in *D. melanogaster* are similar to the Tom element found in *D. ananassae* (53). However, it is also true that a given retrotransposon is usually more closely related to retrotransposons found in extremely diverse species than to many retrotransposons found in the same species. Copia and 1731 are more like Ty1 of the yeast *Saccharomyces cerevisiae* (54), and the plant elements Ta1 (*Arabadopsis thalia;* 55) and Tn1 (tobacco; 56) than they are like any other *Drosophila* transposable element. Conversely, the yeast element Ty3 (57) is more closely related to other *Drosophila* LTR retrotransposons, such as gypsy and 297, than it is to Ty1.

On the whole, this distribution of transposable element families allows two possibilities. It is possible that these families are extremely old, and some members of each group have survived in each species examined through lineal descent. Alternatively, transfer of elements between hosts could occur frequently on an evolutionary time scale. Evidence for the latter hypothesis comes from the case of P elements, which entered the *D. melanogaster* genome only during this century (29,58), and are most closely related to elements found in relatively distant congenic species (59,60). However, the ability of P elements to transfer between species within a genus may not be generalizable to transfer between phyla, and there are other reasons for thinking that this event may be unique (29). Furthermore, although the similarities between (for example) copia and Ty are unmistakable, they are not, in and of themselves, evidence for horizontal transfer, because there are many pairs of homologous yeast and *Drosophila* genes that are more similar.

REGULATION OF ELEMENT ACTIVITY

P/M Hybrid Dysgenesis as the Paradigm

As originally described (61), P elements were the agents responsible for hybrid dysgenesis, a syndrome of traits including high mutability, male recombination, and sterility. These traits were observed in the progeny from crosses between males from P strains (which carry P elements) and females from M strains (which lack them), but not in the reciprocal crosses or in crosses between P strains or between M strains. In addition, the activity of P elements is restricted to the germ line (29), and this has been shown to be because the correct splicing of the messenger RNA (mRNA) for transposase is limited to germ-line tissues (62). The lack of activity in crosses between M strains is easily understood because of the absence of P elements in such cases, but the repression of P element activity in crosses between flies from P strains is a complex phenomenon, which remains incompletely described. The product of the transposase gene in the absence of the germ-line-specific splice does have repressive activity (63), as do other transposase-related products of nonautonomous P elements (64,65). In the laboratory, the activity of P elements is usually controlled by providing transposase for only a single generation (37), bypassing the need for a repressor.

The phenomenon of hybrid dysgenesis has served as the paradigm for an attractive hypothesis about the regulation of transposition in general. According to this model (58,61,66), transposition occurs in bursts when active transposons find themselves in an environment lacking repressor, an indicator of the absence of similar elements. If some of the features of hybrid dysgenesis, particularly the induction of transposition by crosses between strains with active P elements and those without, were general, then the rate of transposition of all elements would be extremely variable, with transposition occurring primarily following crosses between strains with different numbers of the element being mobilized.

I/R and Hobo Hybrid Dysgenesis and Other Examples

In fact, phenomena similar to those seen in hybrid dysgenesis have now been observed for a number of transposable elements, including some examples from three of the four major types just described. An independent system of hybrid dysgenesis brought about by crosses between I (inducer) strain males and R (reactive) strain females was also characterized well before the molecular basis was understood (67). Like P/M hybrid dysgenesis, I/R hybrid dysgenesis has been shown to be the result of the activity of a well-defined transposable element, in this case the I factor, a non-LTR retrotransposon (described above and reviewed in 20). The mobilization of hobo by crosses between H (hobo) and E (empty) strains is more recently described (68,69), but generally is similar to P/M and I/R hybrid dysgenesis (30). Finally, specific strains in which the LTR retrotransposons gypsy (70) and stalker (71) are unusually active have been described. Although the basis of activation (or repression) remains undetermined in these cases, and may differ in significant ways from that seen in hybrid dysgenesis, it is clear that several transposable elements, including examples from most of the major groups, show conditional activity of some sort.

ALTERATION OF GENE EXPRESSION

That an insertion of several kilobases of DNA into a gene would disrupt its function does not seem to be in need of explanation. Indeed, in many cases, transposable element insertions are into coding sequences and simply render the production of functional mRNA impossible. However, a relatively large fraction of mutations caused by the insertion of transposable elements are insertions into noncoding DNA, and many of them result in phenotypes whose severity can be altered by mutations in genetic modifiers or is conditional in some other way. Secondary sequence alterations (such as deletions or further insertions) within a primary insertion often have large effects on the phenotype (72–77), and different transposable elements inserted at identical (or nearly identical) locations can have radically different phenotypes (78,79). Quite often, the phenotypes of insertion mutations are altered by the action of second-site genetic modifiers (reviewed in 9), or can be dependent on nongenetic conditions such as temperature (80). Thus, insertion mutations appear to sensitize gene expression to the action of many factors, both genetic and nongenetic. The importance of this phenomenon is underscored by the observation that insertion mutations account for the majority of spontaneous mutations in *Drosophila* (see 4 and 9) and are responsible for at least some human genetic disease (81). In addition, an insertion is responsible for altered regulation of the mouse C4 variant known as Slp (22). Furthermore, such insertion mutations promise to be very useful in the identification of genes that encode factors with important roles in the molecular processes disrupted by the insertions. I shall consider two examples that are the focus of current research. The allele *yellow-2* (y^2) will illustrate the effects of transposable element insertions on transcription, and *white-apricot* (w^a) will illustrate the effects of transposable element insertions on RNA processing.

Transcription: *suppressor-of-Hairy-wing* and Gypsy

Genetic suppressors in *Drosophila* are mutations that reverse the effect of a mutation mapping elsewhere. In 1943, Lewis (82) observed that mutations in the gene *suppressor-of-Hairy-wing [su(Hw)]* could suppress 15 mutations at 11 loci, all of which were spontaneous. This observation gained new significance in 1983, when Modolell *et al.* (83) discovered that nearly all alleles suppressible by *su(Hw)* are associated with the insertion of the LTR retrotransposon gypsy. The basis of this specificity has been revealed by molecular characterization of both the *su(Hw)* gene and the alleles that are modified by *su(Hw)*. The *su(Hw)* gene encodes a protein structurally similar to transcription factors—it has 12 copies of the zinc-finger motif and an acidic domain (84). Furthermore, the *su(Hw)* protein is capable of specifically binding to sequences within the gypsy element that are related to the octamer motif found in enhancers and promoters of vertebrate genes (85–87), and these sequences have been shown to be responsible for the mutagenic effect of gypsy at *yellow* (88,89) and *bithorax* (90). In the case of the y^2 allele, expression is affected only in those tissues whose enhancers are separated from the site of transcription initiation by the gypsy insertion, and these results are best explained if the *su(Hw)* protein, bound to sites within the gypsy insertion, interferes with the activity of these transcriptional enhancers (91). Thus, one mechanism by which transposable-element insertions can alter gene expression is through interference with the activity of genetic elements, such as transcriptional enhancers, that can act over a distance. The complementary effect, whereby tissue- or stage-specific enhancers can be identified by their effects on the expression of a gene carried within a transposon, is the basis for an important technique for identifying genes with interesting patterns of expression (92–94; Rossant *et al.*, Chapter 4).

RNA Processing: *white-apricot* and *vermillion*

One would expect that insertions within introns would be removed by splicing. Indeed, many insertions within introns are without phenotypic consequence (95). However, some insertions within introns do disrupt gene expression. For example, the *white-apricot* (w^a) allele of *D. melanogaster* is caused by disruption of the 74-bp second *white* intron by the transposable element copia (96–99). Consequently, the level of normally spliced mRNA is greatly reduced, and a number of aberrant RNAs that are polyadenylated within copia are observed (95,73,100,101). Interest in w^a derives from the existence of genetic modifiers that act in trans to alter its expression. Five genetic modifiers are known to act on w^a. These modifiers are specific in their action. Among *white* alleles, only w^a (and some alleles derived from w^a by secondary mutation) are affected by *suppressor-of-white-apricot* [*su*(w^a)] and *suppressor-of-forked* [(*su*(*f*)], whereas *Enhancer-of-white-apricot* [*E*(w^a)] enhances not only w^a (and some of its derivatives), but also w^{a4}, w^e, and w^{sp55} (74,102,103), all of which are associated with the insertion of transposable elements into the transcribed region (98,104). Independent mechanisms of action for these modifiers are implied by the observations that these modifiers are additive in their effects and that no two show identical allele-specificity (95,73,102,105,106).

An important clue to the mechanism of action of *su*(w^a) on w^a has come from a characterization of the expression of the *su*(w^a) gene itself, which has revealed that it is a negative autoregulator acting at the level of splicing (107,108). However, a direct influence of *su*(w^a) on the splicing of *white-apricot* remains to be demonstrated. Because of the competition between RNA splicing events that lead to the production of functional mRNA and polyadenylation events that lead to the production of aberrant transcripts, expression of the w^a allele is likely to be unusually sensitive to mutations that alter RNA processing, and additional modifiers of w^a expression are likely to be alternations in such genes.

Another example in which the expression of an insertion mutation depends on RNA processing is provided by the example of *vermillion* and *suppressor-of-sable*. In this case, the suppressible allele *vermillion-1* is observed to yield functional mRNAs by novel splicing events that imprecisely excise the LTR retrotransposon 412 from the 5′ untranslated region of transcripts from the mutant gene (109). The mechanism of suppression by *suppressor-of-sable* remains unclear, although effects on the efficiency of this splice or on the stability of the resulting RNA are possibilities. This example differs from *white-apricot* in that the splicing events that remove copia sequences from *white-apricot* (and BEL sequences from *white-apricot-4* and blood sequences from *white-blood*) are presumed to occur at the natural splice sites flanking an insertion within an intron, whereas the splicing events that remove 412 sequences from *vermillion* take place at sites provided by the element itself. Similar alleles have been described for a number of plant transposable-element insertions (reviewed in 110). Thus, it appears that transposable-element insertions can alter gene expression by altering the efficiency or regulation of splicing, or even by providing new splice sites.

SUMMARY

Transposable elements are structurally diverse, and each of the major structural groups that can be identified in *D. melanogaster* can also be found in a variety of taxonomically diverse species. This wide distribution of transposable-element families could be due to either extreme age or the transfer of elements between hosts. Insertions of transposable elements can cause mutation not only by disruption of coding regions, but also by interfering with transcriptional enhancers or splicing signals. The expression of genes carrying such insertions often responds to genetic background and environmental conditions in complex ways, and the study of such insertions in *Drosophila* should continue to provide information about both insertional mutagenesis and gene expression in general.

REFERENCES

1. Berg, D. E., Howe, M. M. (eds.). (1989). "Mobile DNA." American Society for Microbiology, Washington, D.C.
2. Lambert, M. E., McDonald, J. F., and Weinstein, I. B. (eds.). (1988). "Eukaryotic Transposable Elements as Mutagenic Agents." Cold Spring Harbor Laboratory, Cold Spring Harbor, New York.
3. Shapiro, J. A. (ed.). (1983). "Mobile Genetic Elements." Academic Press, Orlando, Florida.
4. Finnegan, D. J., and Fawcett, D. H. (1986). *Oxf. Surv. Eukaryotic Genes* **3**, 1–62.
5. Finnegan, D. J. (1989). *Trends Genet.* **5**, 103–107.
6. Boeke, J. D., and Corces, V. G. (1989). *Annu. Rev. Microbiol.* **43**, 403–434.
7. Rubin, G. M. (1983). *In* "Mobile Genetic Elements" (J. A. Shapiro, ed.), pp. 329–361. Academic Press, Orlando, Florida.
8. Xiong, Y., and Eickbush, T. H. (1990). *EMBO J.* **9**, 3353–3362.
9. Bingham, P. M., and Zachar, A. (1989). *In* "Mobile DNA" (D. E. Berg and M. M. Howe, eds.), pp. 485–502. American Society for Microbiology, Washington, D.C.
10. Varmus, H., and Brown, P. (1989). *In* "Mobile DNA" (D. E. Berg and M. M. Howe, eds.), pp. 53–108. American Society for Microbiology, Washington, D.C.

11. Boeke, J. D., Garfinkle, D. J., Styles, C. A., and Fink, G. R. (1985). *Cell* **40,** 491–500.
12. Brown, P. O., Bowerman, B., Varmus, H. E., and Bishop, J. M. (1987). *Cell* **49,** 347–357.
13. Craigie, R., Fujiwara, T., and Bushman, F. (1990). *Cell* **62,** 829–837.
14. Bushman, F. D., Fujiwara, T., and Craigie, R. (1990). *Science* **249,** 1555–1558.
15. Katz, R. A., Merkel, G., Kulkosky, J., Lies, J., and Skalka, A. M. (1990). *Cell* **63,** 87–95.
16. Eichinger, D. J., and Boeke, J. D. (1990). *Genes Dev.* **4,** 324–330.
17. Hutchinson, C. A., III, Hardies, S. C., Loeb, D. D., Shehee, W. R., and Edgell, M. H. (1989). *In* "Mobile DNA" (D. E. Berg and M. M. Howe, eds.), pp. 593–617. American Society for Microbiology, Washington, D.C.
18. Fanning, T. G., and Singer, M. F. (1987). *Biochim. Biophys. Acta* **910,** 203–212.
19. Weiner, A. M., Deininger, P. L., and Efstratiadis, A. (1986). *Annu. Rev. Biochem.* **55,** 631–61.
20. Finnegan, D. J. (1989). *In* "Mobile DNA" (D. E. Berg and M. M. Howe, eds.), pp. 503–517. American Society for Microbiology, Washington, D.C.
21. Finnegan, D. J. (1989). *In* "Mobile DNA" (D. E. Berg and M. M. Howe, eds.), pp. 518–521. American Society for Microbiology, Washington, D.C.
22. Stavenhagen, J. B., and Robins, D. M. *Cell* **55,** 247–254.
23. Xiong, Y., and Eickbush, T. H. (1988). *Mol. Biol. Evol.* **5,** 675–690.
24. Doolittle, R. F., Feng, D.-F., Johnson, M. S., and McClure, M. A. (1989). *Q. Rev. Biol.* **64,** 1–30.
25. Mizrokhi, L. J., Georgieva, S. G., and Ilyin, Y. V. (1988). *Cell* **54,** 685–691.
26. Chaboissier, M.-C., Busseau, I., Prosser, J., Finnegan, D. J., and Bucheton, A. (1990). *EMBO J.* **9,** 3557–3563.
27. Brierley, H. L., and Potter, S. S. (1985). *Nucleic Acids Res.* **13,** 485–500.
28. Hartl, D. L. (1989). *In* "Mobile DNA" (D. E. Berg and M. M. Howe, eds.), pp. 531–536. American Society for Microbiology, Washington, D.C.
29. Engels, W. R. (1989). *In* "Mobile DNA" (D. E. Berg and M. M. Howe, eds.), pp. 437–484. American Society for Microbiology, Washington, D.C.
30. Blackman, R. K., and Gelbart, W. M. (1989). *In* "Mobile DNA" (D. E. Berg and M. M. Howe, eds.), pp. 523–529. American Society for Microbiology, Washington, D.C.
31. Craig, N. L. (1990). *Cell* **62,** 399–402.
32. Engels, W. R., Johnson-Schlitz, D. M., Eggleston, W. B., and Sved, J. (1990). *Cell* **62,** 515–525.
33. Federoff, N. V. (1989). *In* "Mobile DNA" (D. E. Berg and M. M. Howe, eds.), pp. 375–411. American Society for Microbiology, Washington, D.C.
34. Coen, E., Robbins, T. P., Almeida, J., Hudson, A., and Carpenter, R. (1989). *In* "Mobile DNA" (D. E. Berg and M. M. Howe, eds.), pp. 413–436. American Society for Microbiology, Washington, D.C.
35. Spradling, A. C., and Rubin, G. M. (1982). *Science* **218,** 341–347.
36. Rubin, G. M., and Spradling, A. C. (1982). *Science* **218,** 348–353.
37. Robertson, H. M., Preston, C. R., Phillis, R. W., Johnson-Schlitz, D., Benz, W. K., and Engels, W. R. (1988). *Genetics* **118,** 461–470.
38. Engels, W. R., and Preston, C. R. (1981). *Genetics* **26,** 421–428.
39. Geyer, P. K., Richardson, K. L., Corces, V. G., and Green, M. M. (1988). *Proc. Natl. Acad. Sci. U.S.A.* **85,** 6455–6459.
40. Kleckner, N. (1989). *In* "Mobile DNA" (D. E. Berg and M. M. Howe, eds.), pp. 227–268. American Society for Microbiology, Washington, D.C.
41. Campuzano, S., Carramolino, L., Cabrera, C. V., Ruiz-Gromez, M., Villares, R., Boromak, A., and Modolell, J. (1985). *Cell* **40,** 327–338.
42. Potter, S. S. (1982). *Nature (London)* **297,** 201–204.
43. Truett, M. A., Jones, R. S., and Potter, S. S. (1981). *Cell* **24,** 753–763.

44. Green, M. M. (1967). *Genetics* **56,** 467–482.
45. Collins, M., and Rubin, G. M. (1982). *Cell* **30,** 71–79.
46. Bingham, P. M. (1980). *Genetics* **95,** 341–353.
47. Levis, R., Collins, M., and Rubin, G. M. (1982). *Cell* **30,** 551–565.
48. Ising, G., and Block, K. (1981). *Cold Spring Harbor Symp. Quant. Biol.* **45,** 527–549.
49. Goldberg, M. J., Paro, R., and Gehring, W. J. (1982). *EMBO J.* **1,** 93–98.
50. Hoffman-Liebermann, B., Liebermann, D., and Cohen, S. N. (1989). *In* "Mobile DNA" (D. E. Berg and M. M. Howe, eds.), pp. 575–592. American Society for Microbiology, Washington, D.C.
51. Henikoff, S., and Plasterk, R. A. (1988). *Nucleic Acids Res.* **16,** 6234.
52. Fouracade-Peronnet, F., d'Auriol, L., Becker, J., Galibert, F., and Best-Belpomme, M. (1988). *Nucleic Acids Res.* **16,** 6113–6125.
53. Tanda, S., Shrimpton, A. E., Ling-Ling, C., Itayama, H., Matsubayashi, H., Saigo, K., Tobari, Y. N., and Langley, C. H. (1988). *Mol. Gen. Genet.* **214,** 405–411.
54. Mount, S. M., and Rubin, G. M. (1985). *Mol. Cell Biol.* **5,** 1330–1338.
55. Voytas, D. F., and Ausubel, F. M. (1988). *Nature (London)* **336,** 242–244.
56. Grandbastien, M.-A., Spielmann, A., and Caboche, M. (1989). *Nature (London)* **337,** 376–380.
57. Hanson, L. J., Chalker, D. L., Sandmeyer, S. B. (1988). *Mol. Cell. Biol.* **8,** 5245–5256.
58. Bingham, P. M., Kidwell, M. G., and Rubin, G. M. (1982). *Cell* **29,** 995–1004.
59. Lansman, R. A., Shade, R. O., Grigliatti, T. A., and Brock, H. W. (1985). *Proc. Natl. Acad. Sci. U.S.A.* **84,** 6491–6495.
60. Daniels, S. B., Peterson, K. R., Strausbaugh, L. D., Kidwell, M. G., and Chovnick, A. (1990). *Genetics* **124,** 339–355.
61. Rubin, G. M., Kidwell, M. G., and Bingham, P. M. (1982). *Cell* **29,** 987–994.
62. Laski, F. A., Rio, D. C., and Rubin, G. M. (1986). *Cell* **44,** 7–19.
63. Misra, S., and Rio, D. C. (1990). *Cell* **62,** 269–284.
64. Robertson, H. M., and Engels, W. R. (1989). *Genetics* **123,** 815–824.
65. Black, D. M., Jackson, M. S., Kidwell, M. G., and Dover, G. A. (1987). *EMBO J.* **6,** 4125–4135.
66. Engels, W. R. (1981). *Cold Spring Harbor Symp. Quant. Biol.* **45,** 561–565.
67. Bregliano, J. C., and Kidwell, M. G. (1983). *In* "Mobile Genetic Elements." (J. Shapiro, ed.) pp. 363–410. Academic Press, New York.
68. Yannopoulous, G., Stamatis, N., Monastiriorti, M., Hatzopoulos, P., and Louis, C. (1987). *Cell* **49,** 487–495.
69. Blackman, R. K., Grimaila, R., Koehler, M. M. D., and Gelbart, W. M. (1987). *Cell* **49,** 497–505.
70. Mevel-Ninio, M., Mario, M.-C., and Gans, M. (1989). *EMBO J.* **8,** 1549–1558.
71. Georgiev, P. G., Kiselev, S. L., Simonova, O. B., and Gerasimova, T. I. (1990). *EMBO J.* **9,** 2037–2044.
72. Mizrokhi, L. J., Obelenkova, L. A., Priimagi, A. F., Ilyin, Y. V., Gerasimova, T. I., and Georgiev, G. P. (1985). *EMBO J.* **4,** 3781–3787.
73. Mount, S. M., Green, M. M., and Rubin, G. M. (1988). *Genetics* **118,** 221–234.
74. Flavell, A. J., Alphey, L. S., Ross, S. L., and Leigh-Brown, A. (1990). *Mol. Gen. Genet.* **220,** 181–185.
75. Daniels, S. B., McCarron, M., Love, C., and Chovnick, A. (1984). *Genetics* **109,** 95–117.
76. Voelker, R. A., Greenleaf, A. L., Gyurkovics, H., Wisely, G. B., Huang, S.-M., and Searles, L. L. (1984). *Genetics* **107,** 279–294.
77. Geyer, P. K., Green, M. M., and Corces, V. G. (1988). *Proc. Natl. Acad. Sci. U.S.A.* **85,** 3938–3942.
78. Kidd, S., and Young, M. W. (1986). *Nature (London)* **323,** 89–91.

79. Markopoulou, K., Welshons, W. J., and Artavanis-Tsakonas, S. (1989). *Genetics* **122,** 417–428.
80. Bingham, P. M., and Chapman, C. H. (1986). *EMBO J.* **5,** 3343–3351.
81. Kazazian, H. H., Wong, C., Youssoufian, H., Scott, A. F., Phillips, D. G., and Antonarakis, S. E. (1988). *Nature (London)* **332,** 164–166.
82. Lewis, E. B. (1943). *Drosoph. Inf. Serv.* **23,** 59–60.
83. Modolell, J., Bender, W., and Meselson, M. (1983). *Proc. Natl. Acad. Sci. U.S.A.* **80,** 1678–1682.
84. Parkhurst, S. M., Harrison, D. A., Remington, M. P., Spana, C., Kelley, R. L., Coyne, R. S., and Corces, V. C. (1988). *Genes Dev.* **2,** 1205–1215.
85. Spana, C., Harrison, D. A., and Corces, V. C. (1988). *Genes Dev.* **2,** 1414–1423.
86. Mazo, A. M., Mizrokhi, L. J., Karavanov, A. A., Sedkov, Y. A., Krichevskaja, A. A., and Ilyin, Y. V. (1989). *EMBO J.* **8,** 903–911.
87. Spana, C., and Corces, V. C. (1990). *Genes Dev.* **4,** 1505–1515.
88. Geyer, P. K., Richardson, K. L., Corces, V. C., and Green, M. M. (1988). *Proc. Natl. Acad. Sci. U.S.A.* **85,** 6455–6459.
89. Geyer, P. K., Green, M. M., and Corces, V. C. (1990). *EMBO J.* **9,** 2247–2256.
90. Peifer, M., and Bender, W. (1988). *Proc. Natl. Acad. Sci. U.S.A.* **85,** 9650–9654.
91. Harrison, D. A., Geyer, P. K., Spana, C., and Corces, V. C. (1989). *Dev. Genet.* **10,** 239–248.
92. Bier, E., Vaessin, H., Shepherd, S., Lee, K., McCall, K., Barbel, S., Ackerman, L., Carretto, R., Uemura, T., Grell, E., Jan, L. Y., and Jan, Y. N. (1989). *Genes Dev.* **3,** 1273–1287.
93. Bellen, H. J., O'Kane, C. J., Wilson, C., Grossniklaus, U., Pearson, R. K., and Gehring, W. J. (1989). *Genes Dev.* **3,** 1288–1300.
94. Wilson, C., Pearson, R. K., Bellen, H. J., O'Kane, C. J., Grossniklaus, U., and Gehring, W. J. (1989). *Genes Dev.* **3,** 1301–1313.
95. Levis, R., O'Hare, K., and Rubin, G. M. (1984). *Cell* **38,** 471–481.
96. Gehring, W. J., and Paro, R. (1980). *Cell* **19,** 897–904.
97. Bingham, P. M., and Judd, B. H. (1981). *Cell* **25,** 705–711.
98. O'Hare, K., Murphy, C., Levis, R., and Rubin, G. M. (1984). *J. Mol. Biol.* **180,** 437–455.
99. Pepling, M. E., and Mount, S. M. (1990). *Nucleic Acids Res.* **18,** 1633.
100. Pirrotta, V., and Brockl, C. (1984). *EMBO J.* **3,** 563–568.
101. Zachar, Z., Davidson, D., Garza, D., and Bingham, P. M. (1985). *Genetics* **111,** 495–515.
102. Birchler, J. A., and Hiebert, J. C. (1989). *Genetics* **122,** 129–138.
103. Peng, X., and Mount, S. M. (1990). *Genetics* **126,** 1061–1069.
104. Zachar, Z., and Bingham, P. M. (1982). *Cell* **30,** 529–541.
105. Rabinow, L., and Birchler, J. A. (1989). *EMBO J.* **8,** 879–889.
106. Birchler, J., Heibert, J., and Rabinow, L. (1989). *Genes Dev.* **3,** 73–84.
107. Zachar, Z., Chou, T.-B., and Bingham, P. M. (1987). *EMBO J.* **6,** 4105–4111.
108. Chou, T.-B., Zachar, Z., and Bingham, P. M. (1987). *EMBO J.* **6,** 4095–4104.
109. Fridell, R. A., Pret, A.-M., and Searles, L. L. (1990). *Genes Dev.* **4,** 559–566.
110. Wessler, S. R. (1988). *Science* **242,** 399–405.

11

Meiotic Recombination in *Caenorhabditis elegans*

A. M. ROSE AND K. S. McKIM
Department of Medical Genetics
University of British Columbia
Vancouver, B.C., Canada

INTRODUCTION

Meiosis is the process by which eukaryotes reduce their chromosome content by half. During meiosis, the chromosomes undergo two divisions, the first of which involves the organized synapsis of homologous chromosomes. During this division, synapsis predisposes the chromosomes to recombination and proper disjunction. We review here aspects of meiotic recombination under study using the self-fertilizing hermaphroditic nematode *Caenorhabditis elegans*. Six linkage groups were identified by Brenner (1) that correlated with the six chromosomes observed by Nigon (2). Although the behavior of the chromosomes is reported to be holokinetic (3), we have not needed to invoke any unusual mechanisms to explain their behavior with regard to meiotic recombination. On the contrary, there appears to be a single homolog recognition site localized at or near one end of each of the chromosomes. It is not known whether the homolog recognition site is associated with a centromere, but it is clear that this region is responsible for the initiation of the meiotic phenomena of homolog pairing, recombination, and proper disjunction.

HOMOLOG RECOGNITION

Studies of rearrangements have shown that, in *C. elegans*, homolog recognition is determined by localized chromosomal features and not by shared sequence identity.

Mechanisms of Eukaryotic DNA Recombination

Translocations

Rosenbluth and Baillie (4) showed that one portion of the reciprocal translocation, *eT1(III;V)*, paired, recombined, and segregated from only one of two possible "homologs" in a translocation heterozygote. These results showed that only one part of the translocation contained the information required for homolog recognition. McKim, Howell, and Rose (5) reviewed the effects of translocations on recombination frequency. In translocation heterozygotes that have been analyzed, the crossover-suppression boundary and the breakpoint coincide. Recombination is suppressed on only one side of the breakpoint. These observations are most easily explained if the absence of recombination is a consequence of the failure to pair. Thus, sites necessary for homolog recognition must lie outside the crossover-suppressed regions. We have analyzed the behavior of four translocations involving chromosomes I (4,5) and found evidence supporting the hypothesis that sequences at the right end of chromosome I are essential for meiotic pairing and recombination. The translocation *hT2(I;III)* localizes the homolog pairing region to the right third of the chromosome (6).

For other chromosomes in which two or more translocations have been studied, one portion of the chromosome can be identified in which crossing-over is always suppressed. The other portion, which contained sequences necessary for homolog pairing, recombines and segregates from its normal homolog. Three translocations of chromosome III that have been studied, *eT1(III;V)* (4); *sT1(III;X)* (7) and *hT2(I;III)* (5), locate the pairing region in the left portion. The pairing region is in the left portion of chromosome IV, based on studies of *sT2(IV;V)* (7) and *nT1(IV;V)* (8). For chromosome V, the right portion carries the sequences that initiate pairing, as evidenced by studies on *eT1(III;V)* (4), *sT2(IV;V)* (7) *nT1(IV;V)* (8) and *hT1(I;V)* (5). In the case of the X chromosome, studies on *mnT2(II;X)* (9) and *szT1(I;X)* (5) showed that the *unc-1 dpy-3* region paired and recombined with the normal X chromosome, whereas the portion of the chromosome to the right of *dpy-3* was translocated to another chromosome and crossover suppressed.

Duplications

The data described above are in agreement with those from studies using duplications. Although most duplications studied in *C. elegans* do not recombine appreciably with their normal homologs (10), rare duplications of chromosomes I (11), IV (12), and X (13) that do recombine meiotically have been recovered. Each such duplication contains the chromosome region shown to recombine in a translocation heterozygote. For chromosome I, Rose, Baillie, and Curran (11) showed that a duplication of the left portion of

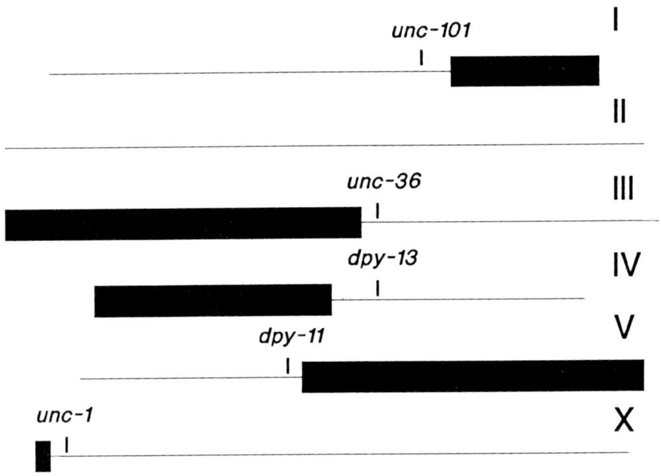

Fig. 1. Representation of the six chromosomes of *C. elegans*. The heavy bars indicate the regions where the homolog recognition sites are located. Data taken from translocation heterozygotes for chromosomes where two or more translocations have been analyzed. See text and ref. 5 for more details.

chromosome I, *sDp2*, did not recombine with the normal homologs, whereas a duplication of the right portion, *sDp1*, did pair and recombine. Studies using duplications of the X chromosome localized the sequences responsible for pairing to the left end of the X chromosome (13). On chromosome IV, the duplication *mnDp1*, which recombines with the normal chromosome IV, includes the left end of the chromosome (12). Thus, chromosomal regions containing homolog recognition sites can be localized by data from studies on both duplications and translocations (Fig. 1).

PAIRING AND RECOMBINATION

Chromosome pairing appears to initiate near one end of the chromosome and continue on the basis of homology. Discontinuity in sequence homology apparently disrupts pairing and consequently recombination.

Translocations

Translocations have been identified on the basis of recombination suppression or pseudolinkage. In those translocation heterozygotes that have been studied, recombination in translocation heterozygotes is suppressed on one side only of the breakpoint (4,5,9). On the other side, recombination occurs up to the breakpoint. It was proposed (5) that pairing was initiated at one end of the chromosome and continued until the heterology was encountered. Recombination is suppressed beyond the breakpoint, because pairing initiation sequences are not present in the translocated portion of the chromosome.

Insertions

Insertions of material have an effect on meiotic recombination. Herman, Albertson, and Brenner (10) observed a polar effect on recombination with the insertional duplication, *mnDp1*. This duplication is inserted into chromosome V, and in heterozygotes recombination is reduced only to the left of the insertion. The insertion of chromosome I material into the X chromosome (*hDp14*) alters X chromosome recombination frequencies in heterozygotes (14). To the left of the insertion site, recombination is either unaffected or slightly increased. To the right of the insertion site, however, recombination is decreased 10-fold (Table I). This suppression of recombination to the right of the insertion agrees with the prediction of our model of homolog recognition in *C. elegans*. The region for homolog recognition on the X chromosome was proposed to be localized at the left end (5). Thus, in the case of the *hDp14* chromosome, the progression of pairing, which had begun at the left end, would have been disrupted by chromosome I sequences. Recombination is not reduced on the proximal side of the insertion because pairing can proceed efficiently up to the insertion site. Similarly for chromosome V, if the right arm initiates the pairing process, then the *mnDp1* situation would be the same as that for *hDp14*. The disruption of recombination, on the side of a heterology that is distal from the homolog recognition site, may be the result of a disruption in pairing. In the cases involving insertion of heterologous material, a reduction in recombination frequency, not elimination, is observed. It is possible that pairing is disrupted by the insertion and poorly reestablished.

TABLE I
Effect of Insertion on Recombination Frequency *(hDp14)*

Chromosomal region	Interval	Recombination frequency[a] (C.I.)[b]	Control	Experimental/ control	Ref.
X left	*dpy-3 unc-20*	6.3(4.6–8.9)	5.0	1.26	11
X right	*dpy-7 unc-3*	1.8(0.9–3.9)	21	0.09	

[a]Expressed as map units.
[b]95% confidence interval.

Other Crossover Suppressors

Herman (15) has described a homozygous viable rearranged chromosome (*mnC1*) that suppresses recombination on the right end of chromosome II. Strains carrying *mnC1* have no other detectable rearrangements. Thus *mnC1* may be an inversion, a deletion that disrupts pairing sequences but does not affect any essential genes, or a DNA insertion. A mutation that eliminates crossing-over for the left half of chromosome I is *hDf8*. This mutation is an embryonic lethal, and complementation testing indicated that there is a deficiency of genetic material (16). In addition, *hDf8* severely reduced recombination over the left half of the chromosome (17).

him-14 Gene

Kemphues, Kusch, and Wolf (18) have described the gene *him-14*. Mutants of this gene were isolated owing to their leaky maternal-effect lethal phenotype; 95% of zygotes from homozygous *him-14* worms were lethal. This lethality was probably the result of extensive meiotic nondisjunction of all chromosomes resulting in extensive aneuploidy. The *him-14* mutants exhibit the phenotype expected, production of few viable progeny, if recombination were absent, and resulted in nondisjunction of all six chromosomes. It is not known whether the *him-14* mutants are defective in homolog pairing or in chiasma formation.

FREQUENCY OF RECOMBINATION

Once pairing has been established, recombination can occur. The frequency of crossing-over is affected by environment, physiology, and mutation.

Temperature and Age

Increasing temperature increased the frequency of recombination (19). In four intervals studied, the recombination frequency at 26°C was twofold higher than that measured at 20°C. Recombination frequency decreased with maternal age (19). In the hermaphrodite, spermatogenesis occurs before and is completed before oogenesis begins (20). This implies that the decrease in recombination frequency with successive broods is a reflection of changing oocyte recombination rate. The oocytes are linearly arrayed in the gonad before reaching the spermatheca where fertilization occurs (21). Temperature upshift experiments indicated a stage in meiosis when

TABLE II
General Increase in Meiotic Recombination Frequency (Rec-1)

Chromosomal region	Interval	Recombination frequency[a] (C.I.)[b]	Control	Experimental/ control	Ref.
Cluster(I)	*dpy-5 dpy-14*	5.5(4.6–6.4)	1.6	3.4	18
Cluster(V)	*dpy-11 unc-42*	7.9(6.1–9.7)	2.2	3.6	18
Noncluster(I)	*bli-3 unc-35*	4.9(4.6–5.2)	1.9	2.6	19

[a]Expressed as map units.
[b]95% confidence interval.

recombination frequency is sensitive to temperature increases (19). The sensitive stage, which was measured at 50 to 75 oocytes before fertilization, corresponds to pachytene. Because recombination frequency in the hermaphrodite changes with temperature and age, standard mapping conditions at 20°C where all progeny are scored has been recommended (19).

rec-1 Gene

Rose and Baillie (22) described the recessive mutation *rec-1(s180)*, which caused a general increase recombination of more than threefold in all regions tested (Table II). Unlike previously described meiotic mutants, the *rec-1* mutation is recessive and leads to an increase in recombination with no detrimental effect on growth rate and displays no elevation in spontaneous mutation rate (23). The *rec-1* gene product appears to function in meiosis and have an effect on flanking marker exchange, conversion, and proper disjunction. It is unlikely that accumulated DNA lesions are the primary cause of the high recombination phenotype because the regional pattern of enhancement is meiosislike, not radiationlike (see the following). Since Rec-1 does not have an increased mutation rate and conversion is increased, it is not likely to be a major repair defect. Rec-1 is not radiation sensitive (24). Both recombination and conversion events are increased. It is as if more heteroduplexes are formed, or more are recombinagenic. The formation of a higher number of crossover events may account for the increased nondisjunction observed in this strain (23).

DISTRIBUTION OF RECOMBINATION EVENTS ALONG THE CHROMOSOME

A number of factors can alter the appearance of the standard genetic map. The examination of some of these factors has indicated that the number of exchange events that occur per unit DNA is regulated during meiosis.

Chromosomal Region

The meiotic map shows a gene-dense region (gene cluster) flanked by gene-sparse regions for each autosome. These clusters are still present on the most recent version of the genetic map (25). Brenner (1) proposed that the clustering resulted from a reduction of recombination frequency per base pair across the clusters, compared to the genome average. Recently, construction of a physical map of the *C. elegans* genome (26,27) has made possible alignment of the genetic and molecular maps for certain regions. Data from these alignments have supported Brenner's proposal (28–30).

Gamma Radiation

Gamma radiation increased recombination frequency across the chromosome I gene cluster, but not outside the cluster (5,31). Ionizing radiation may introduce DNA breaks that can act as substrate for exchange events. Thus, the frequency of recombination would be proportional to the physical distance between markers. The largest increases were observed in the center of the gene cluster, in the *dpy-14 unc-13* interval (Fig. 2). Correlation of the physical map with the genetic map across a portion of the gene cluster demonstrated that the radiation map closely represented the physical distance between markers (30 and Fig. 3).

Gender

The male meiotic map differs from that of the hermaphrodite (32). Several intervals of chromosome I have been studied, and these have either higher, lower, or similar frequencies of recombination, compared with the hermaphrodite. An interval in the chromosome I gene cluster had a reduced frequency of recombination, whereas one at the right end was unchanged.

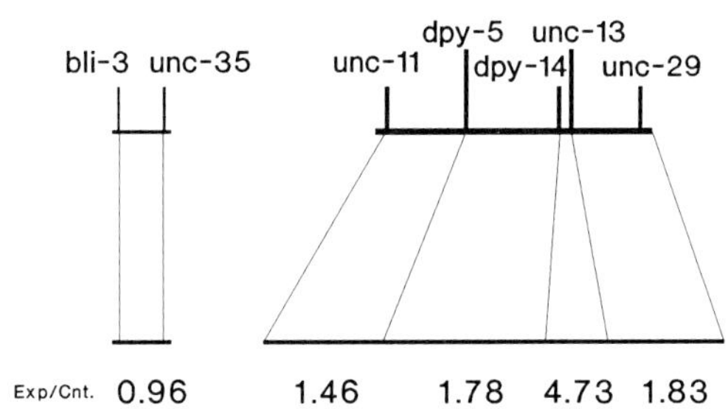

Fig. 2. Radiation map expansion of chromosome I. Top line, standard hermaphrodite map. Bottom line, gamma radiation-induced map. The amount of expansion is shown below the bottom line. Based on data from refs. 31 and 5.

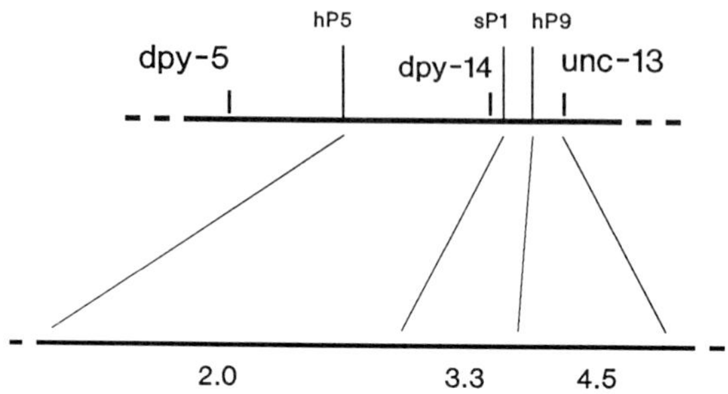

Fig. 3. The relationship between map distance and physical distance for chromosome I. Top line, standard hermaphrodite map. Genetic distances between DNA polymorphisms (*hP5*, *sP1*, and *hP9*) based on data from ref. 30. Bottom line, physical distance between markers estimated from cosmid maps provided by J. Sulston and A. Coulson (MRC, Cambridge, England).

Mechanisms responsible for establishing the pattern of exchange in the male are under investigation.

Rearrangements

Recombination in Nonsuppressed Regions

Translocations and intrachromosomal rearrangements that reduce recombination for portions of the chromosome have an effect on those portions that do pair and recombine. In heterozygotes, translocations of chromosome I have the effect of increasing the level of recombination in the recombining regions of the translocation chromosomes (e.g., *unc-101 unc-54* in Table III) (5, 33). Not only translocations have this property. A deletion of chromosome I that suppressed recombination in the left portion, *h655*, increased the recombination frequency in the right portion (Table III); *mnC1* also has this effect (13), as does *szT1* on the X chromosome (5). The compensatory enhancement of recombination associated with crossover-suppressed chromosomes may be a general phenomenon in *C. elegans* and may reflect a mechanism that is regulating the total number of exchanges per chromosome.

The Translocation, szT1

In the case of *szT1(I;X)*, there is a threefold enhancement of recombination immediately adjacent to the chromosome I breakpoint (5). A second translocation, *hT1(I;V)*, which has a similar breakpoint, does not show any enhancement adjacent to the breakpoint. Furthermore, a III;V translocation, *eT1*, had no detectable increase in recombination adjacent its breakpoints (5). We have hypothesized that *szT1(I;X)* may specifically alter the

TABLE III
Regional Compensatory Increase in Meiotic Recombination Frequency

Genetic background	Interval	Recombination frequency[a] (C.I.)[b]	Control	Experimental/ control	Ref.
szT1	*unc-101 unc-54(I)*	24.4(19.8–29.3)	14.2	1.7	5
hT1	*unc-101 unc-54(I)*	27.0(19.9–35.6)	14.2	1.9	5
h655	*unc-101 unc-54(I)*	23.1(19.6–26.7)	14.2	1.6	14

[a]Expressed as map units.
[b]95% confidence interval.

mechanism that maintains a low level of exchange in the chromosome I gene cluster. The X chromosome does not have a gene cluster, and therefore, similar levels of exchange may occur for all regions of the X chromosome. If the level of exchange is mediated by DNA sequences, then the *szT1* recombination enhancement could be explained by the presence of adjacent X chromosome sequences. The X chromosome sequences might result in higher levels of recombination than would normally be observed across an autosomal gene cluster. On the other hand, the increase in recombination frequency may result from the removal or disruption of chromosome I sequences.

Deletions

Deletions and other intrachromosomal rearrangements suppress recombination in heterozygotes. Deletions along the left half of chromosomes V (34, 35) and I (14) suppress recombination to the right of the regions they delete. The affected region is the opposite end of the chromosome from that identified as important for homolog recognition. On chromosome V, where it was possible to measure recombination on both sides of some deletions, the inhibition was found to be polarized. That is, recombination was normal to the left of the deletion but inhibited to the right along at least 5 map units. These results would not be expected if recombination progressed from the right end, which is believed to initiate pairing. It may be that the left half of LGV is dependent on sites at the right end of the chromosome, which are required for the two homologs to synapse, but that once synapsed, recombination initiates at a number of places (one being the left end of V) and progresses unidirectionally (to the right in this case). One explanation of these data is that the reduction in recombination observed in these studies is a result of a disruption in events that occur after homolog recognition has been established.

DISJUNCTION

Relationship to Recombination

Duplications that recombine tend to segregate from their normal homologs, whereas the nonrecombining duplications do not (11,13). For example, *sDP1/O* gametes are produced from I/I/*sDp1* strains (11). As well, in studies of translocations, it is the portion that pairs and recombines that segregates properly from the normal homolog (4). In cases where this has been studied, the recombining portion segregated from its normal homolog in translocation heterozygotes. The translocated portion, which was recombinationally inert, segregated independent of the normal homolog (4,5,9). Studies in other organisms have implicated chiasma as an essential feature for proper disjunction of homologs (for review, see 36).

The effect of temperature and age on X chromosome nondisjunction has been studied (19). Males (XO) occur in the self-progeny of hermaphrodites as the result of X chromosome nondisjunction (37). At 20°C, the frequency is approximately 1/1000 and increases with increasing temperature (19), possibly as a result of spindle apparatus dysfunction at higher temperatures. Older hermaphrodites produce more male progeny than do younger ones (19). This could be either a consequence of reduced recombination frequency in older hermaphrodites or the direct result of the effect of aging on spindle function.

The *him* Mutants

Hodgkin, Horvitz, and Brenner (37) described 15 mutations of 10 genes that caused increased levels of X chromosome nondisjunction. In the four mutants tested, the nondisjunction occurred at the reductional division, and in three mutants, a large decrease of recombination between X chromosome markers was observed. The analysis of these mutants revealed that some interesting control mechanisms may be operating. Because mutants of at least three of these genes, and probably more, were found to have little effect on the autosomes, it was suggested the X chromosome is partly under independent genetic control. There may be a set of genes that are required only for the X chromosome or there may be different genes for each of the six chromosomes in *C. elegans*. At least two mutants, however, had a phenotype consistent with the hypothesis that they generally increased autosome nondisjunction as well. Two mutants were found to be germ-line specific, based on the production of inviable zygotes. Only when *him-2* was in the sperm line were inviable zygotes produced; *him-7* expressed its phenotype only in the oocytes.

Other Mechanisms of Segregation

Other types of segregation have been observed, so far only in males, that do not depend on recombination. It has been observed that in males, free duplications that do not recombine tend to segregate from the single X chromosome (14,38). Another phenomenon has been observed with *szT1(I;X)*. In heterozygous males it was observed that the half of the translocation that was unpaired in male meiosis tended to segregate from the other translocation chromosome, with which it presumably shared no homology (5).

SUMMARY

Caenorhabditis elegans meiosis is very similar to meiosis in other organisms. We have not needed to invoke any unusual mechanisms to explain the behavior of the chromosomes reported in this chapter, despite their reported holokinetic nature (3).

During *C. elegans* meiosis, a localized region of each chromosome is responsible for homolog pairing. Pairing precedes exchange events, and recombination may be totally absent if homology is lost (in a translocation heterozygote, for example). In other cases in which homology is disrupted (by insertion), recombination may be reduced downstream, relative to the pairing region, of the insertion. The frequency of meiotic exchange can be increased without altering distribution (*rec-1*). The distribution of exchanges along a chromosome is not random and is affected by a number of factors such as chromosomal region, treatment with ionizing radiation, and gender. Chromosomal rearrangements that suppress recombination in one portion of the chromosome may be accompanied by compensatory increases elsewhere. Data from studies involving rearrangements, such as the translocation, *szT1*, suggest that certain DNA sequences may be involved in determination of the frequency of exchange events adjacent to these sequences. Studies on deletions of chromosome V demonstrated a polar reduction in recombination frequency. The polarity of this effect was the opposite of that observed for pairing. The deletions may be interfering with events that occur after homolog pairing is established. Pairing and recombination determine proper segregation of the paired regions. If recombination is reduced or absent, segregation fidelity is reduced.

ACKNOWLEDGMENTS

We thank David Baillie and Raja Rosenbluth for discussion and communication of unpublished data. Supported by a grant from the Natural Science and Engineering Research Council (Canada) to AMR and a Medical Science Council (Canada) studentship to K. S. McK.

REFERENCES

1. Brenner, S. (1974). *Genetics* **77,** 71–94.
2. Nigon, V. (1949). *Ann. Sci. Nat. Zool. Fr.* (ser.11)**2,** 1–132.
3. Albertson, D. G., and Thomson, J. N. (1982). *Chromosoma* **86,** 409–428.
4. Rosenbluth, R. E., and Baillie, D. L. (1981). *Genetics* **99,** 415–428.
5. McKim, K. S., Howell, A. M., and Rose, A. M. (1988). *Genetics* **120,** 987–1001.
6. Peters, K., and K. S. McKim (1989), unpublished results.
7. Rosenbluth, R. E., Cuddeford, C., and Baillie, D. L. (1985). *Genetics* **109,** 493–511.
8. Ferguson, E. L., and Horvitz, H. R. (1985). *Genetics* **110,** 17–72.
9. Herman, R. K., Kari, C. K., and Hartman, P. S. (1982). *Genetics* **102,** 379–400.
10. Herman, R. K., Albertson, D. G., and Brenner S. (1976). *Genetics* **83,** 91–105.
11. Rose, A. M., Baillie, D. L., and Curran, J. (1984). *Mol. Gen. Genet.* **195,** 52–56.
12. Rogalski, T. M., and Riddle, D. R. (1988). *Genetics* **118,** 61–74.
13. Herman, R. K., and Kari, C. K. (1989). *Genetics* **121,** 723–737.
14. McKim, K. (1989), unpublished results.
15. Herman, R. K. (1978). *Genetics* **88,** 49–65.
16. Starr, T. (1989), unpublished results.
17. McKim, K. S., and Rose, A. M. (1989), unpublished results.
18. Kemphues, K. J., Kusch, M., and Wolf, N. (1988). *Genetics* **120,** 977–986.
19. Rose, A. M., and Baillie, D. L. (1979). *Genetics* **92,** 409–418.
20. Hirsch, D., Oppenheim, D., and Klass, M. (1976). *Dev. Biol.* **49,** 200–219.
21. Abi-Rached, M., and Brun, J. L. (1975). *Nematologica* **21,** 151–162.
22. Rose, A. M., and Baillie, D. L. (1979). *Nature (London)* **281,** 599–600.
23. Rattray, B., and Rose, A. M. (1988). *Genet. Res. Camb.* **51,** 89–93.
24. Hartman, P. S., and Herman, R. K. (1985). *Genetics* **109,** 81–93.
25. Edgley, M. L., and Riddle, D. L. (1987). *In* "Genetic Maps" (S. J. O'Brien, ed.), Vol. 4. pp. 351–365. Cold Spring Harbor Laboratory, Cold Spring Harbor, New York.
26. Coulson, A., Sulston, J., Brenner, S., and Karn, J. (1986). *Proc. Natl. Acad. Sci. U.S.A.* **83,** 7821–7825.
27. Coulson, A. Waterston, R., Kiff, J., Sulston, J., and Kohara, Y. (1988). *Nature (London)* **335,** 184–186.
28. Greenwald, I. A., Coulson, A., Sulston, J., and Preiss, L. (1987). *Nucleic Acids Res.* **15,** 2295–2307.
29. Prasad, S., and Baillie, D. L. (1989). *Genomics* **5.** In Press.
30. Starr, T., Howell, A. M., McDowall, J., Peters, K., and Rose, A. M. (1989). *Genome* **32,** 365–372.
31. Kim, J. S., and Rose, A. M. (1987). *Genome* **29,** 457–462.
32. Zetka, M. C., and Rose, A. M. (1990). *Genetics* **126,** 355–363.
33. Peters, K. (1989), unpublished data.
34. Rosenbluth, R. E., Rogalski, T. M., Johnsen, R. C., Addison, L. M., and Baillie, D. L. (1989). *Genet. Res.* **52,** 105–118.
35. Rosenbluth, R. E., Johnsen, R. C., and Baillie, D. L. *Genetics* **124,** 615–625.
36. Hawley, R. S. (1988). *In* "Genetic Recombination" (R. Kucherlapati and G. R. Smith, eds.), pp. 497–527. American Society for Microbiology, Washington, D. C.
37. Hodgkin, J., Horvitz, H. R., and Brenner, S. (1979). *Genetics* **91,** 67–94.
38. Herman, R. K. Madl, J. E., and Kari, C. K. (1979). *Genetics* **92,** 419–435.

PART IV

GENETIC RECOMBINATION IN YEASTS AND *USTILAGO*

12

Saccharomyces cerevisiae Proteins That Promote Hybrid DNA Formation *in Vitro*

RICHARD D. KOLODNER, ERIC ALANI, WOLF D. HEYER[1], ARLEN W. JOHNSON, DAVID N. NORRIS, AND DAN TISHKOFF
Department of Biological Chemistry and Molecular Pharmacology
Harvard Medical School
Boston, Massachusetts 02115
and
Division of Cellular and Molecular Biology
Dana-Farber Cancer Institute
Boston, Massachusetts 02115

RecA-like activities that can form hybrid DNA *in vitro* have been identified in a wide variety of organisms. We have previously described the strand-exchange protein 1 (SEP1) from the yeast *Saccharomyces cerevisiae* that can promote the formation of hybrid DNA *in vitro*. The SEP1 was originally purified as a M_r 132,000 polypeptide; however, recent molecular and immunological studies have now shown that the native form is a M_r 175,000 polypeptide containing both strand-exchange and $5' \rightarrow 3'$-exonuclease activity. The gene encoding SEP1 has been cloned, sequenced, and used to isolate insertion mutations in the *SEP1* gene. Genetic analysis indicated that *SEP1* was essential for normal meiosis, as cells containing a homozygous

1. *Present address:* Institute of General Microbiology, Baltzer-Strasse 4, CH-3012 Bern, Switzerland

Mechanisms of Eukaryotic DNA Recombination

insertion mutation in the *SEP1* gene failed to sporulate. *sep1* Mutants also showed altered meiotic and mitotic recombination properties, had a slow-growth phenotype, and had an altered response to ultraviolet (UV)-irradiation.

In order to identify additional factors that are involved in hybrid DNA formation in *S. cerevisiae*, we used an *in vitro* stimulation of SEP1 assay to identify proteins that reconstitute strand-exchange activity in reactions containing limiting amounts of SEP1. We identified two proteins that functionally interact with SEP1. First, an M_r 34,000 single-stranded, DNA-binding protein stimulated the reaction by lowering the requirement for SEP1 about three- to four-fold. This protein is a fragment of the large subunit of a heterotrimeric complex called yRPA (yRFA), which is thought to be the functional eukaryotic equivalent of prokaryotic single-stranded, DNA-binding proteins. Additional analysis has demonstrated that the intact three-subunit form of yeast RPA will also stimulate SEP1, similar to the M_r 34,000 fragment of the large subunit of RPA. The gene encoding this protein (*RPA1*) is essential for growth. Second, an M_r 33,000 polypeptide, termed Stimulatory *F*actor 1 (SF1), dramatically stimulated the SEP1-catalyzed reaction by lowering the requirement for SEP1 about 300-fold. This protein specifically stimulated SEP1 in a way that is quantitatively and qualitatively different from that observed with other strand-exchange stimulatory proteins.

INTRODUCTION

Genetic recombination is a fundamental cellular process that has been implicated in (1) the generation of genetic diversity, (2) the repair of DNA damage, (3) the homologous alignment of chromosomes required for successful completion of meiotic cell division, and (4) the generation of genomic alterations that lead to changes in gene expression. Using data primarily obtained from tetrad analysis of fungal meiosis, molecular models have been proposed to explain the mechanism of recombination events at the DNA level (1–3). Common to all these models is the central role of the hybrid DNA intermediate. In fungi, hybrid DNA can be directly traced in the phenomenon called PMS (*p*ost *m*eiotic *s*egregation), in which a DNA mismatch originating from the hybrid DNA has escaped mismatch repair, resulting in an impure meiotic product that will segregate during the first postmeiotic mitosis. Thus, PMS can be a direct measure of hybrid DNA formation, giving firm evidence for this intermediate (4).

The identification of the *recA* gene in *Escherichia coli* (reviewed in 5) and the functional analysis of the RecA gene product, further established the pivotal importance of the hybrid DNA intermediate in homologous recom-

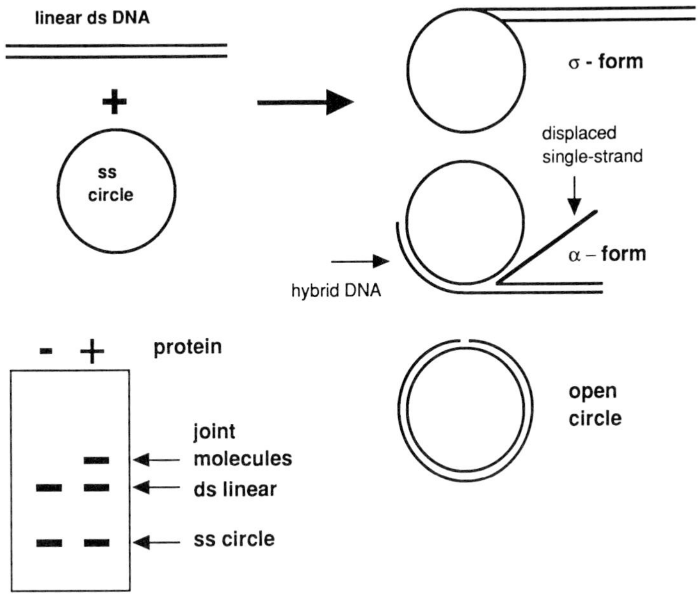

Fig. 1. Hybrid DNA assay: the strand-exchange reaction. Linear double-stranded DNA (M13 replicative form) and circular single-stranded DNA (M13 viral DNA) are the substrates. The products of a strand-exchange reaction are collectively called joint molecules. Electron microscopic analysis can easily distinguish between σ- and α-forms and the end product, open circular DNA. The region of hybrid DNA and the displaced single-stranded region are indicated for the α-form. Also shown is a schematic representation of the migration pattern of the substrates and product molecules in an agarose gel.

bination (reviewed in 6,7). The mechanistics of hybrid DNA formation *in vitro* by the RecA protein has been studied in great detail. Often the model reaction shown in Fig. 1 has been used. The assay makes use of simple DNA substrates that allow detection of the reaction products by gel electrophoresis. The confirmation by electron microscopy of the structure of the joint molecules as drawn in Fig. 1 is essential.

The ease of the strand-exchange assay has prompted the search for RecA-like activities that can form hybrid DNA *in vitro* in many organisms. Such activities have been detected in a wide variety of organisms including fungi (*Ustilago maydis:* 8; *S. cerevisiae:* 9–12), insects (*Drosophila melanogaster:* 13–14), mammals (15–21), and plants (16,22). However, only the fungal activities (cited previously), one protein from *D. melanogaster* (23), and one protein from human cells (24) have been purified to homogeneity. The purified proteins not only allow detailed biochemical characterization of the activity, but also open an avenue to clone and genetically analyze the corresponding genes. The gene–protein sequences will provide valuable information on the phylogenetic relationship of these activities and possibly clues about the structure–function relationship of these proteins. The occurrence of strand-exchange activities in this wide range of organisms suggests some universal mechanisms in the formation of hybrid DNA.

We have used the yeast *S. cerevisiae* as a model system to study the mechanism(s) of homologous recombination using a biochemical approach (25). *Saccharomyces cerevisiae* displays high levels of homologous recombination, and a wealth of data from genetical studies exists (see 4). Together with the experimental possibilities of reverse genetics, yeasts are ideal eukaryotic model systems for use in studying homologous recombination.

Our studies have focused on the mechanism of formation of the central intermediate, hybrid DNA. We have used the assay shown in Fig. 1 to purify and characterize the strand-exchange protein, SEP1 (9). To identify additional proteins involved in hybrid DNA formation, we have used an *in vitro* stimulation assay using limiting amounts of SEP1 to assay for proteins that allow reconstitution of strand-exchange activity. We have identified two proteins that stimulated SEP1 *in vitro:* (1) a single-stranded DNA-binding protein, now known to be the yeast homolog of RPA (yRFA); and (2) a novel protein called stimulatory factor 1 (SF1).

THE STRAND-EXCHANGE PROTEIN, SEP1, FROM *SACCHAROMYCES CEREVISIAE*

Using the assay illustrated in Fig. 1 we have previously purified the strand-exchange protein, SEP1, from mitotic *S. cerevisiae* cells (9). The M_r 132,000 form of SEP1 formed hybrid DNA in the strand-exchange reaction with the polarity shown in Fig. 1, i.e., 3′ to 5′, like *E. coli* recA protein. The SEP1 was required in stoichiometric amounts of one M_r 132,000 monomer per 12 nucleotides of single-stranded DNA in this reaction. The formation of joint molecules required the substrates to be homologous. The presence of either a 4 basepair (bp) 5′ or 3′ single-stranded overhang or blunt end on the linear duplex substrate had no influence on the efficiency of the reaction. The reaction requirements were rather simple; essentially, only the presence of the divalent cation Mg^{2+} was needed. In particular, the strand-exchange reaction occurred in the absence of any requirement for a high-energy cofactor like adenosine triphosphate (ATP). This is in apparent contrast to the situation with RecA, which was thought to require ATP hydrolysis to catalyze strand exchange (7). In addition, SEP1 neither bound ATP, as indicated by photoaffinity-labeling experiments using 8-azido-ATP, nor exhibited ATPase activity [W.-D. H. and R. D. K., unpublished results (1989)]. The absence of an energy requirement does not in principle pose an energetic paradox, as there was no indication for catalytic turnover of the protein. Electron microscopic analysis of the reaction products demonstrated the presence of all three classes of joint molecules as shown in Fig. 1, namely σ- and α-structures as well as the end product, open circular double-stranded DNA. The physical demonstration of α-structures eliminated the possibility that the formation of joint molecules were generated solely by the action of a nuclease followed by reannealing of the resulting homologous single-stranded DNA molecules. The average length of hybrid DNA formed was measured to be 4.1 kbp using a 7.2-kbp substrate.

A number of additional activities are associated with SEP1 that are likely to be important for its action in the formation of hybrid DNA *in vitro* and possibly *in vivo* (26). The SEP1 bound single-stranded DNA as well as double-stranded DNA, but bound double-stranded DNA with lower relative affinity. The binding of SEP1 to the single-stranded DNA rendered the nucleic acid resistant to digestion with nuclease S1 by forming a high-molecular-weight aggregate that effectively precipitated the DNA, apparently excluding access of the nuclease. The formation of these high-molecular-weight aggregates appeared to be a kinetic intermediate in the formation of joint molecules, with strand exchange taking place in these large aggregates. Furthermore, SEP1 catalyzed the renaturation of two complementary single-stranded DNAs, a simple homologous pairing reaction. The mechanism of the renaturation reaction seemed slightly different from that of the strand-exchange reaction in that renaturation was not sensitive to *N*-ethylmaleimide.

These characteristics reveal many similarities to the RecA protein. RecA binds to single-stranded DNA in a cooperative manner and also renatures complementary single-stranded DNA strands, which can occur even in an ATP-independent fashion under certain reaction conditions (as reviewed in 7). In addition, RecA causes DNA aggregation, which is thought to be an intermediate in DNA renaturation and strand exchange (27,28). The difference in the ATP requirement has already been mentioned and will be discussed. Another difference is that SEP1 binds readily to double-stranded DNA, whereas RecA has a kinetic barrier in binding to double-stranded DNA at neutral pH (29).

The gene encoding SEP1 has recently been cloned using degenerate oligonucleotides derived from protein-sequencing data, and the DNA sequence has been established (30,31). The sequence revealed an open reading frame with a coding potential for an M_r 175,000 polypeptide. By searching DNA data bases, *SEP1* was found to be identical to the *KEM1*/*XRN1*/*RAR5* genes identified in other laboratories (32–34); however, the significance of this observation is not yet understood. The full-length M_r 175,000 polypeptide has been overproduced under control of a *GAL10* promoter, and purified to homogeneity (35). It was shown to have potent strand-exchange activity in the strand-exchange assay (see Fig. 1), although it had a slightly higher specific activity than the M_r 132,000 form of SEP1 (35). Analysis of the intact form of SEP1 has indicated that it contains an intrinsic $5' \rightarrow 3'$-exonuclease activity on both single-stranded and double-stranded DNA substrates, in addition to strand-exchange activity (35). Such an exonuclease activity was previously detected in preparations of the M_r 132,000 form of SEP1, but was originally thought to be the result of contaminating exonuclease activity (9). Under appropriate reaction conditions, this exonuclease activity is not required for the strand-exchange activity of SEP1.

The observation of an M_r 175,000 form of SEP1 is in agreement with a recent study of Dykstra *et al.* (12) who have also identified the higher-molecular-weight form of SEP1. Several lines of evidence suggest that the M_r 175,000 polypeptide is the authentic form of SEP1. First, the open reading frame identifying the *SEP1* gene has a coding potential for a polypeptide of this size. Second, NH_2-terminal amino acid sequencing and proteolysis studies confirmed that in the M_r 132,000 form of SEP1, the COOH-terminal ~ 40,000 Da of the M_r 175,000 form were missing. This part of the protein is predicted to be rather disorganized, as it shows a high content of proline residues and was found to be protease sensitive. Third, monoclonal antibodies directed against the M_r 132,000 form of SEP1 detected the M_r 175,000 polypeptide in crude cell extracts.

Monoclonal antibodies directed against SEP1 readily detected the M_r 175,000 polypeptide in crude cell extracts and identified SEP1 as an abundant cellular protein. SEP1 was found to be present throughout the mitotic cell cycle, during meiosis, and was detected in spores. Indirect immunofluorescence light microscopy using anti-SEP1 monoclonal antibodies showed colocalization of SEP1 with the DNA-specific fluorescent probe 4,6-diamidino-2-phenylindole (DAPI), suggesting a nuclear localization of SEP1 [W.-D. H., A. W. J., and R. D. K., unpublished results (1991)].

The analysis (30) of insertion mutations in the *SEP1* gene showed that the gene is not essential for mitotic growth; however, the gene is important for mitotic growth, as the cells carrying the mutation grew more slowly than wild type. *sep1* Mutants were not strongly sensitive to killing by UV-irradiation; however, they showed a greatly increased lag time before initiation of cell division after UV-irradiation compared to wild-type cells. *SEP1* was essential for meiosis since *sep1* mutants were defective in sporulation, resulting in 0 to 3% ascus formation with 60% spore viability. During meiosis *sep1* mutant cells underwent apparently normal premeiotic DNA synthesis, but were blocked before they proceeded through meiosis I. Cells blocked at this stage could be returned to mitotic growth by plating them on appropriate media without significant loss of viability. Examination of the recombination phenotype of *sep1* mutant strains showed a very slight reduction (~ fivefold) in recombination rates between two *his4* heteroalleles during mitotic growth compared to wild type. Induction of recombination during meiosis by a factor of 100 to 1000 is a hallmark of yeast genetics. Induction of meiotic recombination between two *his4* heteroalleles appeared normal in kinetics but as induced to a greater extent in *sep1* mutant strains. When mutant cells were examined for the level of prototroph formation 24 hr after induction of meiosis, they exhibited a six- to tenfold higher meiotic induction than wild-type cells. By this time wild-type cells had already completed meiosis (i.e., sporulated) for about 10 hr. These preliminary data suggest that the SEP1 protein is essential for a successful meiosis, and that the meiotic block in *sep1* mutants occurs at a stage in meiosis at which some type of recombination can take place. An alternate explanation is that *sep1* mutations block a type of recombination required for meiosis, and that the recombination observed in *sep1* mutants is promoted by a parallel pathway.

ADDITIONAL FACTORS INVOLVED IN HYBRID DNA FORMATION *IN VITRO*

A long-term biochemical goal in studying genetic recombination is the reconstitution of large parts or of the entire recombination reaction. In order to identify additional proteins that are potentially involved in hybrid DNA formation, we have purified proteins that stimulate the SEP1 catalyzed reaction *in vitro* under conditions in which the amount of SEP1 was limiting. The results of this *in vitro* stimulation approach are schematically illustrated in Fig. 2. In practice, strand-exchange reactions were set up containing amounts of SEP1 that were too low to sustain activity. Column fractions were systematically assayed for the presence of activities that reconstituted activity. Two proteins were identified and purified using this approach.

M_r 34,000 SINGLE-STRANDED DNA-BINDING PROTEIN [FRAGMENT OF yRPA (yRFA)]

The first stimulatory factor identified was an M_r 34,000 polypeptide that showed nucleic binding properties typical for single-stranded DNA-binding proteins (36). The protein had high affinity for single-stranded DNA, and competition experiments showed that it had no or much lower affinity for double-stranded DNA or RNA. The protein:DNA complexes between the M_r 34,000 polypeptide and single-stranded DNA were highly salt stable (the elution point from single-stranded DNA cellulose was ~ 1.2 *M* NaCl).

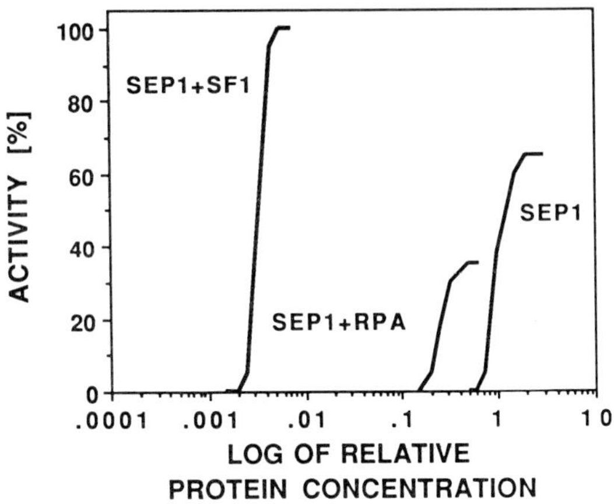

Fig. 2. *In vitro* stimulation of SEP1. Schematic representation of the *in vitro* stimulation of SEP1 by the M_r 34,000 single-stranded DNA-binding protein (fragment of the large subunit of yRFA/RPA) and by the M_r 33,000 polypeptide stimulatory factor 1 (SF1). Activity indicates the percentage of joint molecules formed in the strand-exchange assay described in Fig. 1. The log of relative protein concentration is given to illustrate the amount of stimulation mediated by the different proteins. The amount of protein for SEP1 alone corresponds to 550 m_r 132,000 monomers per single-stranded DNA substrate molecule, for SEP1 + RPA (M_r 34,000 fragment of the large subunit of yRPA or intact yRPA) together corresponds to ~ 150 M_r 132,000 SEP1 monomers per single-stranded DNA substrate molecule, and for SEP1 + SF1 together corresponds to ~ two SEP1 M_r 132,000 SEP1 monomers per single-stranded DNA substrate molecule (30,39).

The M_r 34,000 protein did not have strand-exchange activity by itself. However, when present at an optimal stoichiometry of 1 monomer per 18 nucleotides of single-stranded DNA, the M_r 34,000 protein stimulated the SEP1-catalyzed strand-exchange reaction. As shown schematically in Fig. 2, the amount of SEP1 required to form joint molecules was lowered by three to fourfold. Under these conditions, both the extent and rate of the SEP1-catalyzed reaction were stimulated. *Escherichia coli* SSB could not elicit stimulation of SEP1. Conversely, the M_r 34,000 protein stimulated RecA as well as *E. coli* SSB did. This is consistent with previous observations that RecA can be stimulated by various heterologous single-stranded DNA-binding proteins (37).

The gene encoding the M_r 34,000 protein (*RPA*) was cloned using degenerate oligonucleotide probes and analyzed (38). The open reading frame identifying the gene showed a coding potential for a M_r 70,000 polypeptide. Based on amino acid sequencing data of the M_r 34,000 polypeptide, we deduce that a significant portion of the NH_2 terminus and probably some part of the COOH terminus were missing, probably owing to unspecific proteolysis.

Amino acid sequence homology revealed that the *RPA1* gene encoding this single-strand DNA-binding protein is highly homologous to the large subunit of mammalian RPA (RFA). *R*eplication *p*rotein *A* (RPA), or also called *r*eplication *f*actor *A* (RFA), is a novel type of single-stranded DNA-binding protein identified in higher eukaryotes as a factor essential for *in vitro* SV40 replication and is believed to be the eukaryotic equivalent of prokaryotic SSBs (39–41). The RPA (RFA) is a three-subunit protein with subunits of M_r 70,000, 32,000, and 14,000, which shows high affinity for single-stranded DNA, of which only the M_r 70,000 subunit has DNA-binding activity (42). Recently, a similar trimeric complex has been purified from the yeast *S. cerevisiae* (43).

Because of the relationship between the M_r 34,000 protein and RPA, we have examined the DNA-binding and strand-exchange stimulation properties of the intact trimeric complex purified from yeast. Using fluorescence binding and micrococcal nuclease protection techniques, we determined that yRPA bound to single-stranded DNA as a monomer containing 1:1:1 ratios of the M_r 70,000, 36,000, and 14,000 subunits. Furthermore, yRPA saturated the single-stranded DNA at a stoichiometry of 1 trimeric monomer/100 nucleotides, and binding occurred with high affinity ($> 10^9$ M^{-1}) and cooperativity (ω = 10,000–100,000). Electron microscopic analysis revealed that in yRPA–single-stranded DNA complexes, the single-stranded DNA was compacted fourfold, arranged into nucleosomelike structures, and was free of secondary structure [unpublished results of E. A., R. Thresher, J. Griffith, and R. D. K. (1992)]. Like the M_r 34,000 fragment of the M_r 70,000 DNA-binding subunit, intact yRPA stimulated the strand-exchange activity for both the yeast SEP1 and *E. coli* recA proteins. However, the stoichiometry of intact yRPA required for stimulation of SEP1 was different from that of the M_r 34,000 fragment. The maximal stimulation of SEP1 protein occurred at a stoichiometry of 1 trimeric yRPA monomer/100 bases of single-stranded DNA, as compared to 1 monomer/18 bases for the M_r 34,000 fragment. Electron microscopic analysis of the reaction products of the SEP1-M_r 34,000 fragment and SEP1-RPA-catalyzed reactions demonstrated the occurrence of all three forms of joint molecules, namely σ- and α-structures, as well as open circular end products. Therefore, the reaction products of both the stimulated and unstimulated reactions were indistinguishable in this respect and were formed by a strand-exchange mechanism.

Genetic analysis of insertion mutations in the *S. cerevisiae RPA1* gene revealed that it was essential for mitotic growth (38). Cells lacking *RPA1* arrested as mononucleate, multiply budded structures similar to *cdc* mutations (*cdc4*, 34; as reviewed in 44), believed to be defective in the initiation of chromosomal DNA synthesis. This phenotype is consistent with an involvement of yRPA (yRFA) in DNA replication, as predicted by the requirement for RPA in *in vitro* replication reactions in high eukaryotes. We are presently performing a more detailed genetic analysis of the role of yRPA in both DNA replication and genetic recombination.

STIMULATORY FACTOR 1

A second protein has been identified and purified using this approach. This activity has been called stimulatory factor 1 (SF1), as its presence vastly increased the formation of joint molecules in reactions containing limiting amounts of SEP1 (45). The SF1 was purified using the *in vitro* stimulation assay previously described and was found to copurify with SEP1 over two chromatographic steps, separating during anion-exchange chromatography. As illustrated in Fig. 2, stoichiometric amounts of SF1 (1 M_r 33,000 monomer per 20 nucleotides of single-stranded DNA) lowered the requirement for SEP1 in the strand-exchange reaction to about 1 SEP1 monomer per single-stranded substrate DNA molecule. This represents a roughly 300-fold stimulation as opposed to a three to fourfold stimulation by the M_r 34,000 fragment of RPA1 or intact RPA (see the preceding). This stimulation appeared to be specific to SEP1, as *E. coli* recA protein was not stimulated by SF1. This is in contrast to the stimulation of recA by both the M_r 34,000 fragment of RPA1 and intact RPA. The reaction requirements of the SEP1-SF1-catalyzed reaction remained unchanged and, in particular, the reaction did not require a high-energy cofactor. Electron microscopic analysis of the reaction products of the SEP1-SF1-catalyzed reaction demonstrated the occurrence of all three forms of joint molecules, namely σ- and α-structures, as well as open circular end products. Therefore, the reaction products of both the stimulated and unstimulated reaction were indistinguishable in this respect and were formed by a strand-exchange reaction.

Strand-exchange proteins like SEP1, STPα, or RecA are unspecifically stimulated by conditions that precipitate the substrate DNA, i.e., by the addition of spermidine or basic proteins like histones (10, 12, 45–47). Whereas precipitation of the DNA was certainly one factor in the observed stimulation of SEP1 by SF1, it cannot explain the available data. The SF1 stimulated SEP1 at least one order of magnitude more than other agents that stimulate SEP1 by virtue of their ability to precipitate DNA (addition of spermidine or histones). The effect of SF1 was specific for SEP1, which is in contrast to that of agents like spermidine and histones. Finally, the development of reaction conditions in which the substrate DNA was not precipitated, but a significant stimulatory effect of SF1 was still observed, excluded the possibility that the SF1 effect was entirely due to nonspecific substrate precipitation (45).

The SF1 protein was also characterized in the absence of SEP1 (48). The hydrodynamic properties of the polypeptide indicated that SF1 in solution is a monomer of asymmetric shape (axial ratio 8:1) with a native molecular weight of 31,000. Filter-binding assays showed that SF1 is a DNA-binding protein with preference for single-stranded DNA. Fluorescence-spectroscopic techniques determined the binding-site size of SF1 to be 8 nucleotides of single-stranded DNA and a dissociation constant, k_d, of 2.83×10^{-6} M. The binding to single-stranded DNA appeared noncooperative. In addition, the SF1 polypeptide catalyzed the simplest pairing reaction, the renaturation of two complementary single-stranded DNAs, in the absence of SEP1.

Several different models can explain the mechanism of stimulation of SEP1 by SF1. The SF1 could act directly or indirectly in the strand-exchange reaction. An example of indirect stimulation is the way prokaryotic SSBs stimulate RecA by binding to single-stranded DNA and disrupting inhibitory secondary structure (see 37). The biochemical properties of SF1 and the specificity of stimulation are not compatible with this model. The capability of SF1 to renature single-stranded DNA suggests a more direct involvement of this protein in the strand-exchange reaction. One model suggests a role of SF1 in the initiation of synapsis, with SEP1 promoting branch migration in a subsequent phase of the reaction. This requires that SEP1 molecules recycle within one growing hybrid DNA or recycle to new substrates after completion of one joint, since only one to two SEP1 molecules are required for each joint molecule formed containing several kbp of hybrid DNA. Alternatively, SEP1 might initiate synapsis with SF1 promoting branch migration. In this case, only a small number of SEP1 molecules would be required, whereas a much higher stoichiometry of SF1 would be required. This view most closely reflects the experimental data without invoking recycling of proteins within or between substrates, which we imagine is a process that will require energy, as in the case of RecA (49). A more detailed analysis of the interaction of SEP1 and SF1 has to be awaited to develop more specific models. In addition, the recent cloning of the gene encoding SF1 (unpublished result of D. N. and R. K., 1991) combined with reverse genetics might reveal a more precise role of SF1 in hybrid DNA formation *in vitro* and *in vivo*.

DISCUSSION

A number of proteins exerting strand-exchange activity have been purified from the yeast *S. cerevisiae* (SEP1: 9; STPα: 10; DPA: 11; STPβ: 12), and the question becomes important whether or not they represent different proteins. The apparent molecular weight differences are not reliable for comparison because of the severe proteolysis problems in yeast (i.e., see preceding text for SEP1).

As previously described, SEP1 has been purified from mitotic cells as a M_r 132,000 polypeptide (9). Dykstra *et al.* (12) described the purification of the same activity from mitotic cells as a M_r 180,000 form and renamed it STPβ. The available evidence shows that the two activities are the same protein, with STPβ (M_r of 180,000) possibly being the full-length form of SEP1. The two proteins are both recognized by monoclonal antibodies directed against SEP1. Furthermore, the two proteins show identical proteolytic degradation patterns. Above all, the genes encoding SEP1 and STPβ are the same, based on sequence comparisons (30,31). Given the results of the immunological studies showing that the full-length form of SEP1 appears to be the authentic form in the cell, we conclude that the proteolysis leading to the M_r 132,000 form has been a purification artifact with no likely *in vivo* significance. However, the construction of clones expressing the truncated form of SEP1 and their functional analysis will be required to confirm this view.

The STPα has been purified from meiotic cells and has an apparent M_r of 38,000 (10). In considering the properties of STPα, it is important to note that the strand-exchange activity of STPα does not yet appear to have been demonstrated in the absence of aggregating agents. Immunological studies by Dykstra *et al.* (12) clearly distinguish STPα from SEP1 using polyclonal antisera against the two proteins. Furthermore, they demonstrated the presence of SEP1 protein in a *S. cerevisiae* strain carrying a disruption of the gene *DST1* gene supposed encoding STPα.

Based on protein sequence comparisons, the third strand-exchange activity in *S. cerevisiae*, DPA (11), appears to be distinct from SEP1 (unpublished result of R. D. K. and K. McEntee, 1991).

Therefore, in the yeast *S. cerevisiae* there is good evidence for at least three distinct proteins exerting strand-exchange activity. The existence of redundant activities complicates the interpretation of the phenotype of mutations in any single gene encoding a strand-exchange protein. If the functions of the different proteins are overlapping, a mutational analysis of only one of the genes might not reveal the entire *in vivo* function of the protein. Therefore, the results of the preliminary analysis of insertion mutations in the *SEP1* gene have to be interpreted with this caveat.

All *S. cerevisiae* strand-exchange activities share the characteristic that they do not require a high-energy cofactor to catalyze the reaction *in vitro* (9–11). This property is shared among all purified strand-exchange proteins from eukaryotes (23,24) except for the *Ustilago maydis* Rec1 protein (8). Several activities were found to be dependent on the presence of ATP in crude extract systems or in partially purified fractions (15, 16, 19–21). However, owing to the presence of many proteins in these fractions, it is difficult to assess whether or not the ATP dependence is an intrinsic property of the strand-exchange protein. In fact, an activity from mammalian cells previously described as ATP dependent (21) turned out to be ATP independent on final purification of the protein (24). The observed ATP dependence of the reaction in earlier fractions might have been the result of a limiting stimulating activity that required a high-energy cofactor.

The absence of an energy requirement of many eukaryotic strand-exchange activities was puzzling and in contrast to the concept developed over the years with the *E. coli* RecA protein that strand exchange is an energy-consuming reaction. Initially, it was suggested that RecA protein had to bind and hydrolyze ATP in order to form hybrid DNA (50). Experiments employing the nonhydrolyzable analog of ATP, ATPγS, demonstrated that RecA could bind to DNA and form unstable (paranemic) joints (51). Therefore, it was suggested that the energy-requiring step in the reaction is the conversion from unstable (paranemic) joints to stable (plectonemic) joints. In addition, branch migration was suggested to require ATP hydrolysis (50). In general, the absence of an energy requirement does not pose a paradox, since no catalytic turnover has been demonstrated for any of the ATP-independent strand-exchange protein. In fact, the high stoichiometric amounts of protein needed for the reaction suggest a noncatalytic mode of action. Recently, Kowalczykowski and colleagues (49) made the surprising observation of extended hybrid DNA formation by the RecA protein in the presence of ATPγS, which was hydrolyzed to a minimal extent during the reaction (maximally 0.003 ATPγS molecules hydrolyzed per base pair hybrid DNA formed). They suggested that ATP hydrolysis is required only for product release by the RecA protein. This product-release step could be achieved *in vitro* by the standard treatment of the reaction with sodium dodecyl sulfate/ethylenediaminetetraacetic acid (SDS/EDTA) before gel electrophoresis in the case of the ATP-independent strand-exchange proteins (49). Thus the apparent paradox between the eukaryotic ATP-independent proteins and RecA could be reconciled by suggesting a functional separation of DNA-strand-transfer (ATP not required) and catalytic-turnover (ATP required) activities into physically distinct polypeptides in eukaryotes (49).

ACKNOWLEDGMENTS

This work was supported by the National Institutes of Health Grants GM29383 to R. D. K. W.-D. H. was supported by postdoctoral fellowships from the Swiss National Science Foundation (No. 85BE02) and from the Helen Hay Whitney Foundation (No. F557). A. W. J. was supported by a postdoctoral fellowship from the National Institutes of Health (1F32GM13594-01). D. N. N. was supported by a postdoctoral fellowship from the American Cancer Society

(PF-3218). We thank Dr. A. Sugino for sharing results and reagents that allowed the identification of SEP1 and STPβ as the same protein. We thank Dr. K. McEntee for sharing results to distinguish SEP1 from DPA.

REFERENCES

1. Holliday, R. (1964). *Genet. Res. Camb.* **5,** 282–304.
2. Meselson, M. S., and Radding, C. M. (1975). *Proc. Natl. Acad. Sci. U.S.A.* **72,** 358–361.
3. Szostak, J. W., Orr-Weaver, T. L., Rothstein, R. J., and Stahl, F. W. (1983). *Cell* **33,** 25–35.
4. Orr-Weaver, T. T., and Szostak, J. W. (1985). *Microbiol. Rev.* **49,** 33–58.
5. Smith, G. R. (1988). *Microbiol. Rev.* **52,** 1–28.
6. Radding, C. M. (1982). *Annu. Rev. Genet.* **16,** 405–437.
7. Cox, M. M., and Lehman, I. R. (1987). *Annu. Rev. Biochem.* **56,** 229–262.
8. Kmiec, E. B., and Holloman, W. K. (1982). *Cell* **29,** 367–374.
9. Kolodner, R., Evans, D. H., and Morrison, P. T. (1987). *Proc. Natl. Acad. Sci. U.S.A.* **84,** 5560–5564.
10. Sugino, A., Nitiss, J., and Resnick, M. A. (1988). *Proc. Natl. Acad. Sci. U.S.A.* **85,** 3683–3687.
11. Halbrook, J., and McEntee, K. (1989). *J. Biol. Chem.* **264,** 21,403–21,412.
12. Dykstra, C. C., Hamatake, R. K., and Sugino, A. (1990). *J. Biol. Chem.* **265,** 10,968–10,973.
13. McCarthy, J. G., Sander, M., Lowenhaupt, K., and Rich, A. (1988). *Proc. Natl. Acad. Sci. U.S.A.* **85,** 5854–5858.
14. Eisen, A., and Camerini-Otero, R. D. (1988). *Proc. Natl. Acad. Sci. U.S.A.* **85,** 7481–7485.
15. Kenne, K., and Ljungquist, S. (1984). *Nucleic Acids Res.* **12,** 3057–3068.
16. Hotta, Y., Tabata, S., Bouchard, R. A., Piñon, R., and Stern, H. (1985). *Chromosoma* **93,** 140–151.
17. Hsieh, P., Meyn, M. S., and Camerini-Otero, R. D. (1986). *Cell* **44,** 885–894.
18. Lopez, B., Rousset, S., and Coppey, J. (1987). *Nucleic Acids Res.* **15,** 5643–5655.
19. Ganea, D., Moore, P., Chekuri, L., and Kucherlapati, R. (1987). *Mol. Cell. Biol.* **7,** 3124–3130.
20. Cassuto, E., Lightfoot, L.-A., and Howard-Flanders, P. (1987). *Mol. Gen. Genet.* **208,** 10–14.
21. Fishel, R. A., Detmer, K., and Rich, A. (1988). *Proc. Natl. Acad. Sci. U.S.A.* **85,** 36–40.
22. Peterhans, A., and Paszkowski, J. (1990). "Horticultural Biotechnology." Wiley-Liss, New York.
23. Lowenhaupt, K., Sander, M., Hauser, C., and Rich, A. (1989). *J. Biol. Chem.* **264,** 20,568–20,575.
24. Moore, S. P., and Fishel, R. (1990). *J. Biol. Chem.* **265,** 11,108–11,117.
25. Symington, L. S., Fogarty, L. M., and Kolodner, R. (1983). *Cell* **35,** 805–813.
26. Heyer, W.-D., Evans, D. H., and Kolodner, R. D. (1988). *J. Biol. Chem.* **263,** 15,189–15,195.
27. Tsang, S. S., Chow, S. A., and Radding, C. M. (1985). *Biochemistry* **24,** 3226–3232.
28. Chow, S. A. and Radding, C. M. (1985). *Proc. Natl. Acad. Sci. U.S.A.* **82,** 5646–5650.
29. Pugh, B. F., and Cox, M. M. (1988). *J. Mol. Biol.* **203,** 479–493.
30. Tishkoff, D., Johnson, A. W., and Kolodner, R. (1991). *Mol. Cell. Biol.* **11,** 2593–2608.

31. Dykstra, C. C., Kitada, K., Clark, A. B., Hamatake, R. K., and Sugino, A. (1991). *Mol. Cell. Biol.* **11,** 2583–2592.
32. Kim, J., Ljungdahl, P. O., and Fink, G. R. (1990). *Genetics* **126,** 799–812.
33. Larimer, F. W., and Stevens, A. (1990). *Gene* **95,** 85–90.
34. Kipling, D., Tambini, C., and Kearsey, S. E. (1991). *Nucleic Acids Res.* **19,** 1385–1391.
35. Johnson, A. W., and Kolodner, R. (1991). *J. Biol. Chem.* **266,** 14046–14054.
36. Heyer, W.-D., and Kolodner, R. D. (1989). *Biochemistry* **28,** 2856–2862.
37. Egner, C., Azhderian, E., Tsang, S. S., Radding, C. M., and Chase, J. W. (1987). *J. Bacteriol* **169,** 3422–3428.
38. Heyer, W.-D., Rao, M. R. S., Erdile, L. F., Kelly, T. J., and Kolodner, R. K. (1990). *EMBO J.* **9,** 2321–2329.
39. Wobbe, C. R., Weissbach, L., Borowiec, J. A., Dean, F. B., Murakami, Y., Bullock, P., and Hurwitz, J. (1987). *Proc. Natl. Acad. Sci. U.S.A.* **84,** 18,834–18,838.
40. Fairman, M. P., and Stillman, B. (1988). *EMBO J.* **7,** 1211–1218.
41. Wold, M. S., and Kelly, T. J. (1988). *Proc. Natl. Acad. Sci. U.S.A.* **85,** 2523–2527.
42. Wold, M. S., Weinberg, D. H., Virshup, D. M., Li, J. J., and Kelly, T. J. (1989). *J. Biol. Chem.* **264,** 2801–2809.
43. Brill, S. J., and Stillman, B. (1989). *Nature (London)* **342,** 92–95.
44. Pringle, J. R., and Hartwell, L. H. (1981). *In* "The Molecular Biology of the Yeast Saccharomyces. Life Cycle and Inheritance." J. N. Strathern, E. W. Jones, and J. R. Broach, eds. pp. 97–142. CSH, Cold Spring Harbor Laboratory, New York.
45. Norris, D. N., and Kolodner, R. D. (1990a). *Biochemistry* **29,** 7903–7911.
46. Hamatake, R. K., Dykstra, C. C., and Sugino, A. (1989). *J. Biol. Chem.* **264,** 13,336–13,342.
47. Shibata, T., DasGupta, C., Cunningham, R. P., and Radding, C. M. (1980). *Proc. Natl. Acad. Sci. U.S.A.* **77,** 2606–2610.
48. Norris, D. N., and Kolodner, R. D. (1990b). *Biochemistry* **29,** 7911–7917.
49. Menetski, J. P., Bear, D. G., and Kowalczykowski, S. C. (1990). *Proc. Natl. Acad. Sci. U.S.A.* **87,** 21–25.
50. Cox, M. M., and Lehman, I. R. (1981). *Proc. Natl. Acad. Sci. U.S.A.* **78,** 3433–3437.
51. Riddles, P. W., and Lehman, I. R. (1985). *J. Biol. Chem.* **260,** 170–173.

13

Gap Repair *in Vitro* Catalyzed by Cell-Free Extracts from Yeast

LORRAINE S. SYMINGTON
Institute of Cancer Research and Department of Microbiology
Columbia University College of Physicians and Surgeons
New York, New York 10032

INTRODUCTION

Genetic studies of meiotic recombination in fungi have contributed significantly to our understanding of the mechanisms of recombination. Recombination events may be reciprocal or nonreciprocal (gene conversions). Meiotic gene conversion at a heterozygous locus is detected as a departure from 2:2 segregation, resulting in either a 3:1 or a 1:3 tetrad. Since conversion events are frequently associated with reciprocal exchange (1), it has been suggested that gene conversion is an intermediate in the formation of a reciprocal exchange (2). Several models have been proposed to explain the molecular mechanisms of recombination (reviewed in 3). The models propose that conversion can occur by mismatch repair of a heteroduplex intermediate (4, 5), or by repair of a double-strand gap in one duplex, using the homologous duplex as a template (6). The models share several common features, but differ by the mechanism of initiation.

Several lines of evidence suggest that double-strand breaks either initiate recombination, or provide efficient substrates for recombination. First, DNA-damaging agents such as radiation or chemical carcinogens increase the frequency of mitotic recombination (7). Several DNA-repair pathways have been identified in yeast by mutational analysis (for a review, see 8). These include excision repair, error-prone repair, photoreactivation, and recombinational repair. Mutants of genes in the *RAD50* epistasis group (*rad24,50–57*) were isolated on the basis of sensitivity to ionizing radiation and most have defects in recombination (9). Mutants of *RAD51*, *RAD52*, *RAD54*, and *RAD57* are extremely sensitive to ionizing radiation, almost completely defective in meiotic and some mitotic recombination events, and are impaired in the repair of DNA double-stranded breaks (9–12). Second, mating-type interconversion is initiated by a double-stranded break at the *MAT* locus (13). The HO endonuclease cleaves a specific site at *MAT* and initiates the replacement of these sequences, using one of the donor cassettes, *HMR* or *HML*, as a template for conversion (14). Mating-type switching requires the products of *RAD51*, *RAD52*, *RAD54*, and *RAD57* (11, 15, Hoekstra and Heffron, personal communication). Double-stranded breaks generated by the HO endonuclease can stimulate both mitotic and meiotic recombination when the HO-cut site is placed in novel locations (16–19). Third, Orr-Weaver *et al.* (20) have shown that linear nonreplicating plasmids containing a double-stranded break in yeast DNA sequences transform at much higher frequency than uncut plasmids. Transformants contain plasmid DNA integrated into the yeast genome at the homologous site. Plasmids containing gaps also transform with high efficiency and are repaired by recombination using chromosomal information as a template (20). In these experiments, the duplex containing the break is the recipient of genetic information. Transformation with linear integrating plasmids requires *RAD52* and is partially dependent on *RAD1*, one of the genes involved in the UV excision-repair pathway (20, 21). During meiosis, recombination occurs at frequencies several orders of magnitude greater than that observed during mitotic growth. The frequency of meiotic gene conversion at different loci varies by as much as 40-fold (2). At the *ARG4* locus, alleles at the 5′ end of the gene convert eightfold more frequently than alleles at the 3′ end of the gene; a phenomenon known as polarity (22, 23). When alleles at the low end of the gene convert, they generally coconvert with alleles at the high end of the gene. These observations suggest that there is a site required for the initiation of gene conversion at the high end of the gene. A putative initiation site was mapped by deletion analysis to a region containing the promoter of the *ARG4* gene (23). Interestingly, a meiosis-specific double-strand break was found within this region (24). There is also evidence for meiosis-specific double-strand breaks within an insert of the *LEU2* gene adjacent to the *HIS4* gene; this region was shown genetically to have high frequencies of crossing-over (Cao and Kleckner, personal communication).

Since double-strand breaks are efficiently utilized by the cellular recombination apparatus, we have examined the effects of breaks in an *in vitro* recombination system. Previous experiments have shown that homologous plasmids undergo recombination *in vitro* catalyzed by yeast cell-free extracts (25). Intact plasmids, or plasmids containing double-stranded breaks, were used as substrates. Since the substrates were completely homologous, except for mutant alleles within the Tc gene, the donor and recipient molecules were indistinguishable in these experiments. This system has been refined by using partially homologous substrates. Our results show that the substrate containing the double-stranded lesion is the recipient of information and that a double-stranded break is required within the region of homology between the two substrates.

RESULTS AND DISCUSSION

It has been shown that gapped plasmids are efficiently repaired when transformed into yeast by a mechanism that involves recombination (20). The following substrates and assays were designed to detect recombinational repair of a double-stranded gap *in vitro*, and to distinguish between donor and recipient molecules. The substrates for this assay were derived from the plasmids pACYC184 (26) and pBR322 (27); they share 1.64 kb of homology through the Tc^r gene. The remainder of the plasmid sequences are nonhomologous and contain different drug-resistance markers. The plasmid pLS89 was constructed by deleting a 14-bp fragment between two *Ban*II sites present within the coding region of the Tc gene of pBR322 (Fig. 1) This plasmid is Tc^s and contains a single *Ban*II site. Plasmid pLS96 was generated by deleting 90 base pairs (bp) between the *Cla*I and *Xba*I sites of pACYC184 to delete the promoter of the Tc gene; this substrate is also Tc^s (Fig. 1). The end point of this deletion extends into the region of nonhomology between the two substrates. pLS96 has to be the donor of genetic information for these events since the end point of the promoter deletion is outside of the region of homology, and thus a functional Tc^r gene cannot be restored in pLS96 by homologous recombination. A double-stranded gap can be introduced into pLS89 by cleavage with *Ban*II, which cuts at the site of the deletion. This linear substrate is used in combination with supercoiled pLS96 to detect activities in extracts that can repair the break to generate a Tc^r gene.

Since intrachromosomal events generally occur at higher frequency than interchromosomal events, an additional substrate has been constructed to study this reaction. pLS108 was constructed by first inserting an *Xba*I linker into the Ban II site of pLS89, then joining this plasmid to pLS96. This heterodimer plasmid contains both mutant Tc genes in direct orientation and contains a single restriction-enzyme site within one of the repeats (Fig. 1).

Yeast cell-free extracts are prepared by the method of Symington *et al.* (25). Briefly, haploid yeast strains are grown to early/mid log phase in rich medium (YEPD: 1% yeast extract; 2% Bacto-peptone; 2% glucose). Cultures

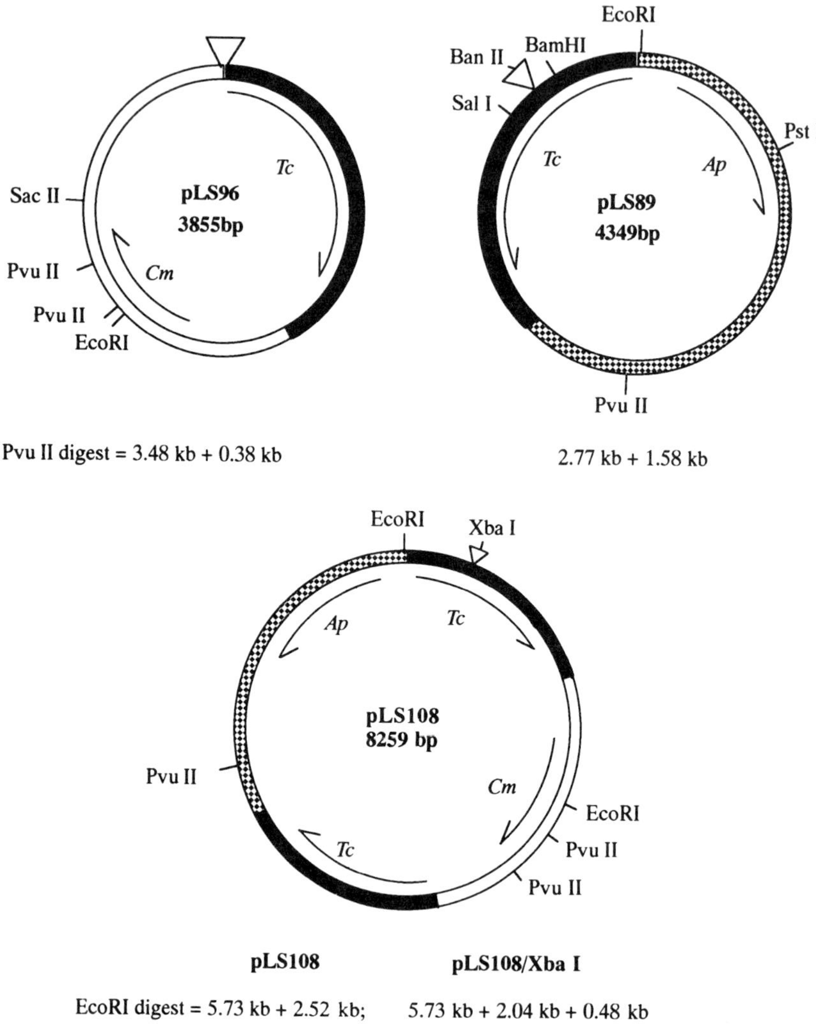

Fig. 1. Substrates for *in vitro* gap repair. The homologous regions between the two plasmids are shown in solid lines. The deletion present in each plasmid is shown by a triangle (see text for details). The sizes of the fragments generated following cleavage with *Pvu*II or *Eco*RI are shown.

are harvested, frozen in liquid N_2, thawed on ice, spheroplasted with zymolyase, and then lysed with a gentle detergent. Following a high-speed clearing spin, proteins are precipitated from the soluble fraction by the addition of saturated $(NH_4)_2SO_4$ to 60% saturation. The protein pellets obtained by this procedure can be stored for up to 1 month at −70°C without significant loss of activity.

Two assays have been used to detect recombinational repair of the double-stranded break in pLS89. The first involves transforming the substrates, after incubation with the extract, into an *Escherichia coli recA* strain (DH5α) and measuring the frequency of Tc^r relative to Ap^r and Cm^r. This provides a convenient and sensitive measurement of the recombination frequency. The second assay detects repair directly by Southern analysis of the reaction products. For this assay, the products are cleaved with either *Pvu*II or *Eco*-RI, and the resulting fragments are separated by agarose gel electrophoresis and Southern blotted. The fragments obtained are unique for each substrate. The probe used for hybridizations is an oligonucleotide that has the same sequence as the gap in pLS89. The oligonucleotide probe efficiently hybridizes to pLS96, but not to pLS89. If repair of the gap occurs, pLS89 fragments will hybridize to the oligonucleotide probe.

Using the transformation assay we have found that the extract stimulates the formation of Tc^r molecules by 20- to 50-fold when a double-stranded break is made within the region of homology (Table I). A gap of 262 base pairs (bp) generated by cleaving pLS89 with *Bam*HI and *Sal*I, was also repaired, but less efficiently than the 14-bp gap. Digestion of pLS89 with *Pst*I to create a double-stranded break within a region of nonhomology does not give rise to Tc^r transformants (Table I). When the two substrates are incubated

TABLE I
Gap Repair Dependent on Cleavage within a Region of Homology[a]

Substrates	Incubation time (min)	Tc^r/Ap^r
pLS89/*Ban*II + pLS96	90	5.65×10^{-2}
pLS89/*Bam*HI/*Sal*I + pLS96	90	2.05×10^{-2}
pLS89/*Pst*I + pLS96	90	$<1 \times 10^{-3}$
pLS89/*Ban*II	90	$<5 \times 10^{-4}$
pLS96	90	$<1 \times 10^{-6}$
pLS89/*Ban*II + pLS96	0	3.04×10^{-4}

[a]Extracts were prepared from strain LBL1N (α, *ade5*, *met13-c*, *cyh2*^{r}, *trp5*, *ade6*, *cly8*, *lys2-2*, *tyr 1-2*, *his7-1*, *ade2-1*, *ura3-1*) as described. Reactions contained 1 μg of each substrate; 35 m*M* *N*-2-hydroxyethylpiperaeine-*N*^{1}-2-ethanesulfonic acid (Hepes), pH 7.8; 10 m*M* $MgCl_2$; 5 m*M* ATP; 1m*M* spermidine, 1 m*M* nicotinamide adenine dinucleotide (NAD); 50 μ*M* each guanosine triphosphate (GTP), cytidine triphosphate (CTP), and uridine triphosphate (UTP); 10 μ*M* each dNTP, 2 m*M* dithiothreitol (DTT); 100 μg/ml bovine serum albumin (BSA) and 0.2–0.4mg/ml yeast extract protein in 50μl. Reactions were performed for the designated time, stopped by the addition of EDTA to 10 m*M*, phenol extracted and ethanol precipitated. DNA was resuspended in 30 μl of TE, and 5 μl used for each transformation.

TABLE II
Homologous Interaction between the Substrates in the Extract is Required for Repair

Substrates	Incubation time (min)	Tc^r/Ap^r	Cm^{r}[a]
pLS89/*Ban*II + pLS96	90	8.7×10^{-2}	42.5
pLS89/*Ban*II + pLS96 + *Sac*II[b]	90	7.5×10^{-2}	3.3
pLS89/*Ban*II + pLS96	90[c]	2.9×10^{-3}	N.D.[d]
pLS89/*Ban*II + pLS96 + *Sac*II	90[c]	$<3.1 \times 10^{-3}$	N.D.
pLS89/*Ban*II + pLS96	0	1.76×10^{-4}	N.D.

[a] Tc^r transformants were replica plated to Cm containing LB plates to determine the percentage that were Cm^r.

[b] *In vitro* reactions were digested with *Sac*II before transformation.

[c] The two substrates, pLS89/*Ban*II and pLS96, were treated independently with extract. The DNA was purified from each reaction and then mixed before transformation.

[d] N.D., Not determined.

independently in the extract, and then mixed before transformation, the frequency of Tc^r molecules is 10-fold higher than that in the substrates that have not been treated with the extract (Table II). This control suggests that treatment with the extract renders the substrates more recombinogenic in the *E. coli recA* strain. However, the stimulation observed is much lower than that observed when the substrates are incubated together in the extract. All of the Tc^r transformants obtained are Ap^r, and about 43% are also Cm^r. No Cm^r, Tc^r, Ap^s transformants were detected. Analysis of the DNA isolated from these transformants has shown that of the Cm^r, Tc^r, Ap^r transformants, 80% consist of both monomer plasmids, and 20% have a single dimer-sized plasmid (data not shown). Since it is possible that these events are owing to *recA* independent recombination following cotransformation of these substrates, we have treated reacted DNA with *Sac*II, which cleaves pLS96, but not pLS89. This should eliminate the chances of cotransformation of the substrates. After this treatment the number of Tc^r transformants decreased by only 14%, and now only 3% were Cm^r. These results show that pLS96 is the donor of information in yeast extract-catalyzed events.

A time course of the reaction between pLS96 and pLS89/*Ban*II is shown in Fig. 2. Figure 2a shows an ethidium bromide-stained agarose gel of reactions incubated from 5 to 90 min, and then digested with *Pvu*II. The donor plasmid, pLS96, yields fragments of 3.48 kb and 0.38 kb; pLS89/*Ban*-II yields fragments of 2.77 kb and 1.58 kb. Figure 2b shows the Southern hybridization of the same gel. Hybridization to the brands of 2.77 kb and 1.58 kb is diagnostic of repair of the gap in pLS89. An increase in the amount of hybridization of the oligonucleotide probe to the fragments derived from

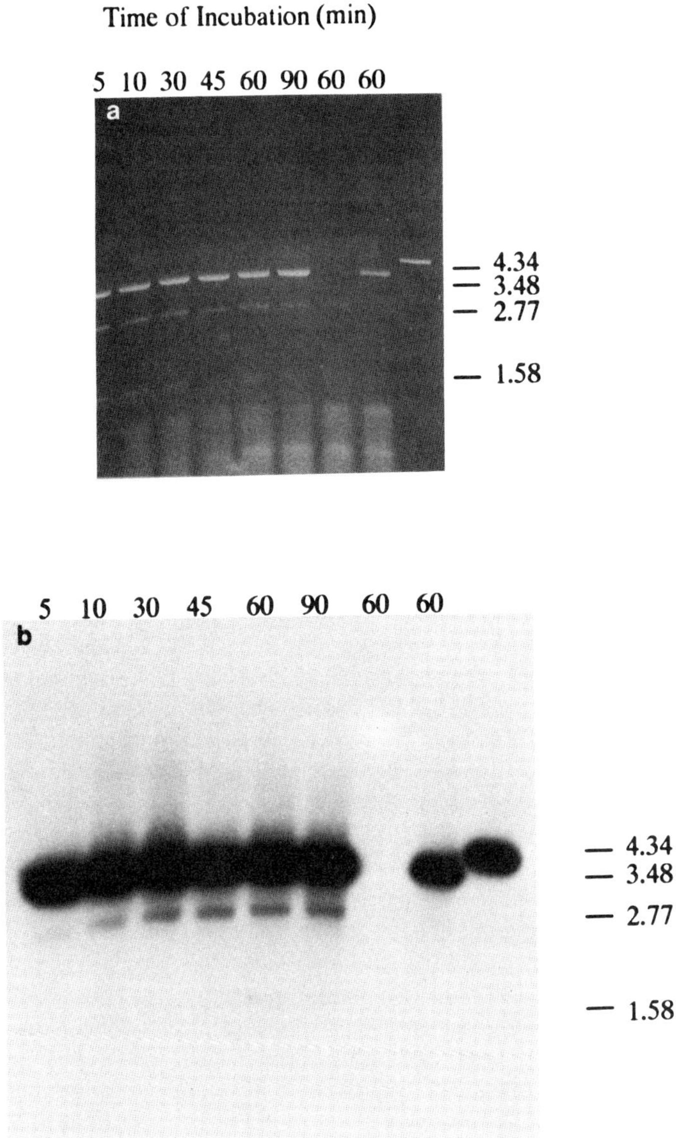

Fig. 2. Southern blotting assay for gap repair. (a) Reactions cleaved with *Pvu*II before agarose gel electrophoresis (0.8% agarose, stained with ethidium bromide). DNA was transferred from the gel to Hybond-N membrane (Amersham Corporation, Illinois) using a Stratavac vacuum blotting apparatus (Stratagene, California). The filter was hybridized with 5′-[γ^{32}P] ATP-labeled 24-base oligonucleotide in 10% (w/v) polyethylene glycol (PEG) 8000; 6 × SSPE; 0.5% (w/v) sodium dodecyl sulfate (SDS); 5 × Denhardt's solution for 4 hr. The filter was washed as follows: 2 × SSC at room temperature for 5 min, repeated once; 2 × SSC, 0.1% SDS at the T_m of the oligonucleotide (75°C) for 10 min, repeated once. The film was exposed for 90 min at −70°C with intensifying screens.

pLS89 is observed through time, reaching a plateau between 60 and 90 min. At later time points, extensive degradation of the substrates occurs (data not shown). The controls for this experiment are the two substrates incubated independently in the extract for 60 min. By Southern analysis there are no bands corresponding to the repaired products presents in either of the controls. These results indicate that a homologous interaction between the substrates in the extract is required for the formation of repaired products. The broken ends of pLS89 can presumably pair and invade the donor plasmid. These ends can serve as primers for DNA synthesis, which generate sequences homologous, and therefore able to hybridize, to the oligonucleotide. Although these results indicate that several steps toward generating a repaired product proceed, a complete repair event should generate a fragment of the same size as linear pBR322 (conversion end product), or fragments of 5 kb and 3 kb (reciprocal exchange end products). These are detected at very low frequency. A model for how this repair pathway may occur is shown in Fig. 3. This is derived from the double-strand break repair model for recombination (6). A number of different enzymatic activities are expected to participate in this recombination pathway. For example, nucleases may be required to generate single-stranded tails for invasion into the homologous duplex; strand-exchange proteins (SEPs) and single-strand DNA-binding proteins (SSBs) for pairing and strand invasion; DNA polymerase to carry out repair synthesis, thus repairing the gap; and DNA ligase to seal nicks. An intermediate in this reaction should be a figure-eight intermediate in which the substrates are joined together by two Holliday junctions. Intermediates with a figure-eight structure have been detected between homologous circular plasmids (28). Since the substrates used in the current experiments are only partially homologous, cleavage with enzymes that cut within the regions of nonhomology will generate X-forms. These will be stable since they cannot branch migrate apart. X-form intermediates will have different mobility than the substrates when electrophoresed through agarose gels, migrating in a similar position to linear dimers. Molecules of this size have been detected on long exposures of the filter shown in Fig. 2b. We have also recently shown that these molecules can also be detected by two-dimensional (2D) agarose gel electrophoresis using techniques similar to those of Bell and Byers (29) and Brewer and Fangman (30) for separating branched molecules (data not shown).

The heterodimer plasmid is a very efficient substrate for these reactions. Figure 4 shows time course reactions of the linear heterodimer substrate, followed by cleavage with *Eco*RI before gel electrophoresis. Digestion of pLS108/*Xba* I with *Eco*RI generates three fragments of 5.73, 2.04, and 0.47 kb; the largest band hybridizes to the oligo probe. The two smaller bands will also hybridize to the oligo probe if repair of the deletion mutation

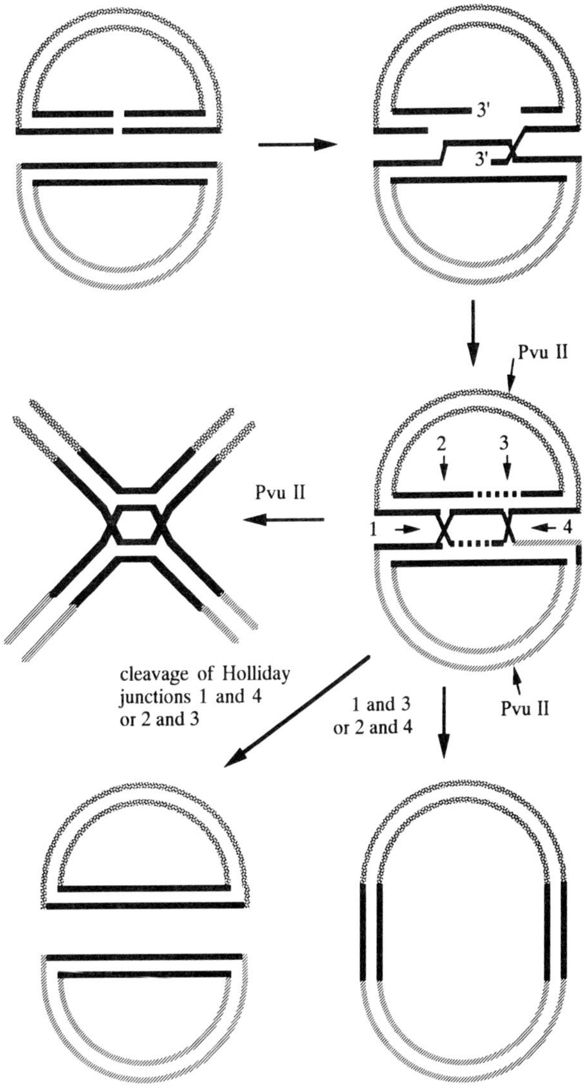

Fig. 3. Model for recombination between plasmids by double-strand break repair (6). The gapped plasmid is degraded by nucleases to generate 3′ ends. The 3′ end invades the homologous duplex and primes DNA synthesis, repairing one strand of the gap. The other 3′ end of the gapped plasmid pairs with the displaced strand and primes DNA synthesis of the other strand. By ligating the nicks that remain following DNA synthesis, a double Holliday structure is generated. If this structure is digested with *Pvu*II a stable X-form is produced. Cleavage of the Holliday junctions present in the figure-8 intermediate yields crossover (dimer) or noncrossover (monomers) products.

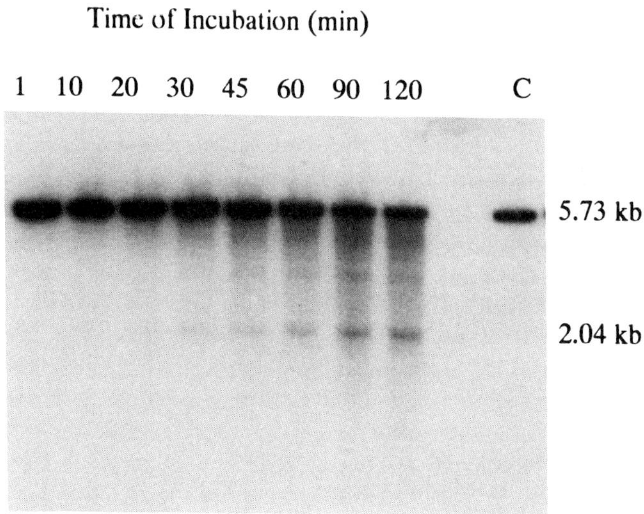

Fig. 4. Time course of repair of heterodimer plasmid. The autoradiogram of reactions containing pLS108/*Xba*I, followed by cleavage with *Eco*RI before gel electrophoresis is shown. Southern blotting and hybridizations were performed as described in the legend to Fig. 2.

occurs. With increasing time of incubation in the yeast extract, the 2.3-kb fragment hybridizes to the oligonucleotide probe. We have constructed a series of isogenic strains containing *rad* disruption or deletion mutations (primarily of the *RAD52* series). In addition, we have prepared extracts from isogenic *rad* mutant strains with a different genetic background (MacDonald and Rothstein, personal communication) to verify that differences observed between mutants and wild type are mutant specific rather than strain specific. Preliminary results have shown that some of the *rad* mutants have wild-type activity using these assays (*rad 51*, *rad54*), whereas others have decreased activity (*rad1*, *rad52*, and *rad55*).

In summary, we have detected recombinational repair of a double-stranded break using two different assays, transformation and Southern blotting. We have shown that intermolecular and intramolecular recombination occurs efficiently and is dependent on cleavage within a region of homology. The duplex containing the break is the recipient of genetic information. For this reaction we presume that degradation of ends, homologous pairing, strand invasion, and DNA repair synthesis are required. To obtain a complete recombination reaction, DNA ligase and a Holliday junction endonuclease are also required. Using the Southern blotting assay we see little evidence for the production of complete recombinants. Although the transformation assay gives rise to Tc^r recombinants, it is possible that the final steps of this reaction occur in *E. coli*. Preliminary results have shown that intermediates in this reaction can be detected by 2D agarose gel electrophoresis. By combining these assays, in conjunction with the use of *rad* mutant strains, we hope to be able to elucidate the biochemical roles of the *RAD* gene products in the recombinational repair of double-stranded breaks.

ACKNOWLEDGMENTS

I should like to thank A. Rattray for critical reading of this manuscript. I should also like to thank G. Cole, M. Esposito, J. MacDonald, and R. Rothstein for gifts of plasmids and strains. This work was supported by NIH Grant GM41784-01 and ACS Grant ACS-IN-177.

REFERENCES

1. Hurst, D., Fogel, S., and Mortimer, R. (1972). *Proc. Natl. Acad. Sci. U.S.A.* **69,** 101–105.
2. Fogel, S., Mortimer, R., Lusnak, K., and Tavares, F. (1979). *Cold Spring Harbor Symp. Quant Biol.* **43,** 1325–1341.
3. Orr-Weaver, T., and Szostak, J. (1985). *Microbiol. Rev.* **49,** 33–58.
4. Holliday, R. (1964). *Genet. Res. Camb.* **5,** 282–304.
5. Meselson, M., and Radding, C. (1975). *Proc. Natl. Acad. Sci. U.S.A.* **72,** 358–361.
6. Szostak, J., Orr-Weaver, T., Rothstein, R., and Stahl, F. (1983). Cell **33,** 25–35.
7. Haynes, R., and Kunz, B. (1981). *In* "The Molecular Biology of the Yeast Saccharomyces" (J. Strathern, E. Jones, and J. Broach eds.), Vol. 1, pp. 371–414. Cold Spring Harbor Laboratory Press, Cold Spring Harbor, New York.
8. Friedberg, E. (1988). *Microbiol. Rev.* **52,** 70–102.
9. Game, J. (1983). *In* "Yeast Genetics: Fundamental and Applied Aspects" (J. Spencer, D. Spencer, and A. Smith eds.) pp. 109–137. Springer-Verlag, New York.
10. Resnick, M., and Martin, P. (1976). *Mol. Gen. Genet.* **143,** 119–129.
11. Malone, R., and Esposito, R. (1980). *Proc. Natl. Acad. Sci. U.S.A.* **77,** 503–507.
12. Cole, G., Schild, D., Lovett, S., and Mortimer, R. (1987). *Mol. Cell. Biol.* **7,** 1078–1084.
13. Strathern, J., Klar, A., Hicks, J., Abraham, J., Ivy, J., Masmyth, K., and McGill, C. (1982). *Cell* **31,** 183–192.
14. Kostriken, R., Strathern, J., Klar, A., Hicks, J., and Heffron, F. (1983). *Cell* **35,** 167–174.
15. Weiffenbach, B., and Haber, J. (1981). *Mol. Cell. Biol.* **1,** 522–534.
16. Kolodkin, A., Klar, A., and Stahl, F. (1986). *Cell* **46,** 733–740.
17. Nickoloff, J., Chen, E., and Heffron, F. (1986). *Proc. Natl. Acad. Sci. U.S.A.* **83,** 7831–7835.
18. Nickoloff, J., Singer, J., Hoekstra, M., and Heffron, F. (1989). *J. Mol. Biol.* **207,** 527–541.
19. Rudin, N., and Haber, J. (1988). *Mol. Cell. Biol.* **8,** 3918–3928.
20. Orr-Weaver, T., Szostak, J., and Rothstein, R. (1981). *Proc. Natl. Acad. Sci. U.S.A.* **78,** 6354–6358.
21. Shiestl, R., and Prakash, S. (1988). *Mol. Cell. Biol.* **8,** 3619–3626.
22. Fogel, S., Mortimer, R., and Lusnak, K. (1981). *In* "The Molecular Biology of the Yeast Saccharomyces" (J. Strathern, E. Jones, and J. Broach, eds.). Vol. 1, pp. 289–339. Cold Spring Harbor Laboratory Press, Cold Spring Harbor, New York.
23. Nicolas, A., Treco, D., Schultes, N., and Szostak, J. (1989). *Nature (London)* **338,** 35–39.
24. Sun, H., Treco, D., and Szostak, J. (1989). *Nature (London)* **338,** 87–90.

25. Symington, L., Fogarty, L., and Kolodner, R. (1983). *Cell* **35,** 805–813.
26. Chang, A., and Cohen, S. (1978). *J. Bacteriol* **134,** 1141–1156.
27. Bolivar, F., Rodrigues, R., Betlach, M., and Boyer, H. (1977). *Gene* **2,** 75–93.
28. Symington, L., Morrison, P., and Kolodner, R. (1985). *Mol. Cell. Biol.* **5,** 2361–2368.
29. Bell, L., and Byers, B. (1983). *Anal. Biochem.* **130,** 527–535.
30. Brewer, B., and Fangman, W. (1987). *Cell* **51,** 463–471.

14

Meiotic Recombination in *Schizosaccharomyces pombe*: Genes, Enzymes, and Sites

GERALD R. SMITH AND ALFRED S. PONTICELLI[1]
Fred Hutchinson Cancer Research Center
Seattle, Washington 98104
and
Department of Pathology
University of Washington
Seattle, Washington 98195

INTRODUCTION

Genetic recombination appears to play at least two important roles during meiosis. The proper pairing and disjunction of homologous chromosomes frequently fail when recombination is blocked; the apposition of chromosomes for recombination may provide the opportunity for their proper segregation, or vice versa (see chapter 9, in this volume). Perhaps secondary to this crucial role for recombination, the exchange of genetic information during recombination generates chromosomes with new combinations of alleles, which in the next generation produce new phenotypic combinations on which natural selection can act. The generally higher frequencies of recombination during meiosis than during mitosis indicate that at least one rate-limiting component of recombination is critically altered during meiosis. Identifying this component, and others, will lead to an understanding of the molecular basis of recombination in meiosis and perhaps in mitosis as well. The similarity of the overall process of meiosis in all eukaryotes suggests that the underlying mechanism of recombination may be similar in organisms as diverse as humans and fungi.

1. *Present address:* Department of Biological Chemistry and Molecular Pharmocology, Harvard Medical School, Boston, Massachusetts 02115

The unicellular fission yeast *Schizosaccharomyces pombe* is particularly well suited for studies of the genetic and enzymatic basis of recombination. During the past 40 years, more than 450 nuclear genes have been identified by mutations and mapped (1). Comparisons between mutant *S. pombe* strains is facilitated by their isogenicity: the commonly used strains are derived from the single homothallic (selfmating) strain chosen by Leupold in 1947 (2). Hundreds of millions of meiotic products (spores) are easily produced and analyzed. Dissection and analysis of the four spores resulting from a single meiosis, initially held in an ascus, permit the distinction of reciprocal recombination (crossing-over) from nonreciprocal recombination (gene conversion). Large cultures of cells ($> 10^{10}$) proceeding synchronously through meiosis are easily prepared by shifting the cells to nitrogen-free medium; even more convenient is the thermal induction of meiosis in a *pat1* (parthenogenic) temperature-sensitive mutant (3). Active enzymes can be readily extracted and purified from these meiotic cells. Finally, a chromosomal site (*M26*) that enhances meiotic recombination has been described in an easily analyzed gene *ade6* (4); analysis of *M26* may help elucidate the genetic control of recombination by special sites, which appear to be widespread among both eukaryotes and prokaryotes (5).

GENES

Among *S. pombe* mutants previously surveyed for meiotic recombination, only one (*swi5*) has an altered frequency of interchromosomal recombination: it is reduced by about 10-fold (6,7). In our laboratory we developed a rapid screen for meiotic recombination proficiency and isolated mutants in which the recombinant frequency is reduced (8,9). We anticipate that the genes thus identified encode products required for recombination.

Our screen for meiotic recombination-deficient mutants incorporates several features that permit the facile isolation of recessive, as well as dominant, mutations. The parent strain is a homothallic (h^{90}) haploid; that is, during mitotic growth, the cells switch their mating type so that they can mate with each other. Thus, a mutation induced in the parent strain will be homozygous in the diploids produced from these matings. The genetic heterozygosity necessary to monitor recombination is provided by an *ade6* mutation on the chromosome and another, noncomplementing *ade6* mutation in the *ade6* gene borne on a multicopy plasmid. In this way, the parent strain is haploid except for the *ade6* gene. *ade6* mutants accumulate a purine biosynthetic precursor that spontaneously polymerizes into a nondiffusible red pigment (10); therefore, *ade6* mutants form red colonies, whereas *ade*$^+$ cells form white colonies. We chose the *ade6-M26* mutation on the chromosome because it has a 15-fold higher frequency of intragenic *ade6* recombination than other *ade6* alleles; this choice may allow the isolation of mutants in which the *M26* hot spot does not act (4; see the following).

The procedure to isolate mutants is simple. The parent strain GP66 [h^{90} *ade6-M26 ura4-294* (pade6-L469)] is mutagenized with nitrosoguanidine and plated for isolated colonies on a solid minimal medium (plus adenine) supporting mitotic growth and mating-type switching. When the colonies become large and nutrients are depleted, cells of opposite mating type within each colony mate and sporulate. About 6 days after plating the cells, each colony contains about 10^6 spores, each resulting from a meiosis homozygous for any mutation induced in the parent strain. During meiosis, recombination between the plasmid and the chromosome can generate *ade*$^+$ cells. The colonies with their spores are transferred with toothpicks to the wells of a microtiter dish containing glusulase to liberate the spores from their asci and to kill vegetative cells. The spore suspensions are spotted onto a rich medium, on which they germinate and grow into red lawns with white papillae from the *ade*$^+$ recombinants, and onto a minimal medium, on which only Ade$^+$ colonies can grow. Clones giving fewer Ade$^+$ papillae and fewer Ade$^+$ colonies than the parent strain are purified and tested in several ways to distinguish genuine recombination-deficient (Rec$^-$) mutants from other types of mutants and variants. For promising candidates, heterothallic (h^+ and h^-) derivatives coupled with appropriate *ade6* (or other) mutations are constructed by standard crosses. A key criterion for our Rec$^-$ mutants is a reduced frequency of recombination between chromosomal alleles of a test gene (e.g., *ade6*) in standard heterothallic matings. The *rec* mutants described here have this property.

In our first search, we obtained 13 *bona fide rec* mutants from screening about 5000 mutagenized cells (8), and a second search produced 40 mutants from about 18,000 mutagenized cells (9). Each Rec$^-$ character appears to be the result of a single nuclear mutation [8,9; N. Hollingsworth, personal communication (1988)]. Fifty-two of the mutations are recessive and define 16 complementation groups. When homozygous, these mutations reduce *ade6* interchromosomal recombination between 3- and 1000-fold. One mutation analyzed further is dominant and, whether heterozygous or homozygous, reduces *ade6* interchromosomal recombination about fourfold. The 16 *rec* genes appear to fall into three discrete classes with respect to the residual level of recombination in the mutants (9). Mutations in the class I genes reduce recombination about 1000-fold and appear to leave only the low mitotic level of recombination. Mutations in the class II and class III genes reduce recombination about 100- and 5-fold, respectively. Our laboratory has isolated multiple alleles of the six class I genes and of the three class II genes; the mutant isolation may, therefore, be nearly saturated for this type of gene. Only single alleles of the seven class III genes have been isolated.

As expected from the view that recombination is essential for proper meiotic chromosome disjunction, the fraction of the meiotic spores that are viable is related to the residual level of recombination in each *rec* mutant (8). Spores from mutants with slightly reduced levels of recombination are essentially all viable, but only about 15% of the spores from the most recombination-deficient mutants are viable. Since *S. pombe* has only three chromosomes, and aneuploids grow poorly or not at all (1,11), random segregation of the chromosomes in meiosis is expected to produce viable (euploid) spores at a frequency of 2^{-3} (12.5%), close to the observed value for the most recombination-deficient mutants. Our results are therefore consistent with the hypothesis that recombination is essential for orderly chromosome segregation during meiosis.

Our laboratory has begun a genetic characterization of the *rec* mutations. Mutations in the eight *rec* genes tested to date have no significant effect on mitotic recombination between chromosome- and plasmid-borne *ade6* alleles or between two chromosomal *ade6* or two *ura4* alleles [K. Larson, N. Hollingsworth, and J. Virgin, personal communications (1990; 1988; 1991)]. Mutations in only three of the sixteen *rec* genes (*rec9*, *rec17*, and *rec19*) confer detectable sensitivity to DNA-damaging agents (methylmethane sulfonate or UV light) during mitotic growth (8,9). The *rec* mutations analyzed to date therefore appear to be, in the main, meiotic specific. Although the *ade6-M26* allele was incorporated in the screen with the expectation of isolating mutations specifically reducing the *M26* stimulation (described in succeeding sections), none of the *rec* mutations representing the 16 complementation groups appears to do so (8,9). Fourteen *rec* genes have been assigned to chromosomes by mitotic chromosome loss, and three *rec* mutations have been mapped more precisely: *rec6* is about 2 centimorgans (cM) from *pat1* in the order *cenII-rec6-tps13-pat1* [P. Szankasi, personal communication (1989)], *rec11* is linked to *ade6*, and *rec15* is linked to *mat* (9). The other *rec* mutations are not closely linked to *ade6*, *mat*, or *ura4*. The *rec* mutations therefore appear to inactivate diffusible products.

Our laboratory has developed a procedure for isolating *rec* gene clones from existing *S. pombe* genomic, plasmid-borne libraries. A given *rec* mutant, containing appropriate mutations for selecting recombinants, is transformed with the library and mated with the same *rec* mutant containing appropriate complementary mutations. Prototrophic recombinants are selected, and plasmid DNA is recovered in *Escherichia coli*. The *rec* mutant is transformed and retested for recombination-proficiency. With this procedure the *rec7* and *rec8* genes have been cloned and sequenced (Lin, Y., Larson, K., Dorer, R., and Smith, G., manuscript in preparation). Analysis of these and other *rec* genes and their products may provide insight into the biochemical activities of the *rec* genes.

The presumed products of the *rec* genes appear to have differential effects

TABLE I
Differential Reduction of Meiotic Recombinant Frequencies by *S. pombe rec* Mutations[a]

	Relative recombinant frequency at		
rec	*ade6*	*ura4*	*pro2-arg3*
+	1	1	1
rec6	0.001	<0.05	0.1
rec7	0.001	<0.01	0.03
rec8	0.001	<0.1	1.0
rec9	0.3	0.2	0.3
rec10	0.01	0.1	0.5
rec11	0.003	0.1	0.2
rec12	0.001	<0.05	0.1
rec13	0.1	0.2	0.3
rec14	0.001	<0.1	0.2

[a]Standard heterothallic matings homozygous for a mutation at the indicated *rec* gene were conducted. Prototrophic spore frequencies, among total viable spores, were determined by differential platings. In the *rec*$^+$ crosses, the frequency of Ade$^+$ was 3.0×10^{-3}; of Ura$^+$, 3.6×10^{-4}; and of Pro$^+$ Arg$^+$, 1.2×10^{-1}. The alleles crossed were *ade6-M26* × *ade6-L52*, *ura4-294* × *ura4-595*, and *pro2-1* × *arg3-124*. The data are summarized from ref. 8 and 9.

at different test loci. For example, the *rec11-111* mutation reduces intragenic *ade6* recombination about 300-fold, intragenic *ura4* recombination about 10-fold, and intergenic *pro2 - arg3* recombination about 5-fold (Table I). The *rec10-109* mutation reduces *ade6* recombination 100-fold, but reduces *pro2 - arg3* recombination only 2-fold. The *rec8-110* mutation reduces *ade6* recombination 1000-fold but has little, if any, effect on *pro2 - arg3* recombination.

There are several possible interpretations of these differential effects. Recombination in one test interval might be more sensitive to mismatch correction than that in another; the nucleotide sequences of the alleles at each test interval are likely to be different, and mismatches of different sequences may be corrected with different efficiencies. In addition, mismatch correction may more severely influence recombination of close markers (intragenic) than recombination of distant markers (intergenic). A similar differential effect could result from altered lengths of hybrid DNA in the mutants. Alternatively, the *rec* gene products may be more directly involved in recombination within some test intervals than within others; for example, recombination at *ade6* may require the *rec10* and *rec11* gene products more than does recombination at *ura4*. Genes controlling recombination in a region-specific fashion have been described in *Neurospora crassa* (12). Finally, some *rec* gene products may be required for intragenic recombination (gene conversion) but not for intergenic recombination (crossing-over). Further analysis may establish the regional, allelic or gone conversion specificity of the *S. pombe rec* gene products.

ENZYMES

Studies of recombination in prokaryotes, principally *E. coli* and its phages, have shown that homologous recombination proceeds by the breakage and joining of DNA molecules, in some cases with associated DNA synthesis (13). Identified enzymatic activities required for prokaryotic recombination include those of nucleases, strand-transfer activities, those of single-stranded DNA-binding proteins, DNA unwinding activities, those of DNA polymerases and DNA ligases, and activities that cleave branched DNA structures such as cruciforms and Holliday junctions (13). In eukaryotes, principally *Saccharomyces cerevisiae* and humans, some of these activities have been detected, and the enzymes, purified; these include strand-transfer activities, single-stranded DNA-binding proteins, DNA ligase, and cruciform-cleaving activities (14), and see chapters by Kolodner and by Symington in this volume.

Three of these enzymatic activities have been detected in extracts of *S. pombe* [15 and P. Szankasi, T. Shibata, R. Steinglanz personal communications (1991)]. An activity that pairs homologous circular single-stranded DNA and linear double-stranded DNA was detected using substrates frequently used for strand-transfer assays (described in ref. 13). An activity that cleaves a cruciform substrate (Holliday junction-resolving enzyme) was also detected. Two nuclease activities have been detected and purified to near homogeneity. Exonuclease I (Exo I) activity is induced about fivefold during meiosis. Like phage lambda exonuclease, Exo I digests double-stranded linear DNA from the 5′ ends and produces mononucleotides plus linear molecules with single-stranded tails having 3′ ends (15). There is strong evidence that such a structure is an intermediate in prokaryotic recombination (13), and it has been hypothesized as an intermediate in fungal recombination (16, 17). Exonuclease II (Exo II), like the *E. coli* RecJ exonuclease, digests single-stranded DNA from the 5′ end (P. Szankasi, and G. Smith, 1992). This activity may, in conjunction with a helicase, generate 3′ single-stranded tails or, in conjunction with an endonuclease, digest D-loops formed by homologous pairing of single-stranded and double-stranded DNA. Using the partial amino acid sequences of the purified proteins, the genes for Exo I and Exo II have been isolated [P. Szankasi, personal communication (1991)]. The behavior of mutants lacking Exo I or Exo II will reveal their role, if any, in recombination.

SITES

Special sites controlling meiotic recombination have been identified in several fungal species (5). These sites are presumably recognition sites for proteins that either promote or repress recombination in a region-specific manner. In *S. pombe* Gutz (4) isolated the *ade6-M26* mutant as an adenine auxotroph and showed that the mutation created a site that enhanced the frequency of conversion within the *ade6* gene. In comparison with the *ade6-M375* mutation, which maps very close to *ade6-M26* and converts at about the same frequency as other ("ordinary") *ade6* alleles, *ade6-M26* converts and recombines with other *ade6* alleles at about 10 to 15 times higher frequency. *ade6-M26* stimulates conversion of other *ade6* alleles, both to its right and left. Unlike other *ade6* alleles, which show parity in the direction of conversion, *M26* is preferentially converted to wild type (i.e., *M26* is a recipient of genetic information). These properties led Gutz (4) to propose that *M26* creates a site recognized by an endonuclease; after the DNA is cut at or near *M26*, an exonuclease degrades DNA to the right or to the left (or both), and replacement synthesis using the homolog as a template leads to conversion of *M26* and flanking alleles. We and others have begun a genetic and physical analysis of *M26* to test this hypothesis.

Nucleotide sequence analysis revealed that *ade6-M26* is a single base-pair (bp) change from wild type: G:C → T:A (18,19). The wild-type and several *ade6* mutant genes were cloned and sequenced. Comparisons showed that *ade6-M26* creates a 5′ T-G-A 3′ codon [5′ U-G-A 3′ in messenger RNA (mRNA)], as expected from the nonsense suppressibility of the adenine auxotrophy (4). (The recombination stimulation of *M26* is not suppressed.) *ade6-M375* is three nucleotides away from *M26* and also changes 5′ G-G-A 3′ to 5′ T-G-A 3′ (see Fig. 1). *ade6-M375* thus is an excellent control for studies of *M26*. Although the T:A bp created by *M26* is essential for its action, other nucleotides must also be important. Surrounding the *M26* site, there are no striking symmetries or repeats. Site-directed mutagenesis has revealed that a continuous stretch of seven nucleotides (5′ A-T-G-A-C-G-T 3′) is essential for hot-spot activity: any change of any of these seven nucleotides virtually abolishes *M26* activity (20). These results support the hypothesis that *M26* is a protein-recognition site. An M26-specific binding protein has been detected in extracts of *S. pombe* cells. [W. Wahls; E. Käslin and J. Kohli, personal communications (1990)].

Genetic experiments have revealed additional properties of *M26*. *M26*

```
5' ATT GAT GG[A] G[GA CGT] GAG 3'
           ↓     ↓
           T    [T]
        M375    M26
```

Fig. 1. Nucleotide sequence surrounding the *M26* hotspot. The upper line shows part of the nucleotide sequence of the wild-type *ade6* gene (18). The single base-pair changes of the *ade6-M375* and *ade6-M26* mutations are indicated (18,19). The outlined nucleotides, or their complement, or both, are required for *M26* hotspot activity (20).

stimulates recombination at *ade6* but not at *arg1*, located about 80 cM to the right of *ade6* on chromosome III; *M26* has no significant effect on intergenic recombination in three separate intervals on chromosomes I and II (19). These results show, as expected, that *M26* acts by creating a site that locally stimulates recombination and not by altering the product of the *ade6* gene. A mismatch at the *M26* site is not required for its action; i.e., even when homozygous, *M26* stimulates recombination of other *ade6* alleles (Table II). Therefore, *M26* stimulation does not simply result from biased mismatch correction at the *M26* site, although mismatch correction at the recombining alleles may be involved. *M26* stimulates meiotic, but not mitotic recombination. This feature has been demonstrated for standard chromosome-by-chromosome recombination (Table III), as well as chromosome-by-multicopy plasmid recombination (Table III) and recombination between directly repeated DNA (21). These results suggest that the hypothetical *M26*-recognizing protein is induced or activated during meiosis or that the chromatin structure at *M26* is changed during meiosis. Collectively, the preceding results demonstrate that *M26* is a meiotic recombination-promoting site.

Although *M26* contains a 7 bp (base pair) protein-binding site, additional experiments have shown that it either is part of a large complex site or

TABLE II
Action of *M26* When Homozygous[a]

Cross at *ade6*	Parent with *M26*	Ade^+ per 10^6 viable spores
	Neither	150
(*M26*) *L52* +	Top	1300
(*M26*) + *M210*	Bottom	260
	Both	1700

[a]Standard heterothallic crosses were analyzed for Ade^+ spores by differential plating. Each parent contained the *sup9* UGA-suppressor, which suppresses the adenine auxotrophy of the *ade6-M26* mutation (4). The *ade6-M26* mutation was present in the parents as indicated. The data are from 19.

TABLE III
Hot Spot Activity of *M26*[a]

		Ade$^+$ per 10^6 mitotic cells or meiotic spores	
Cross		Mitotic	Meiotic
Chromosome × chromosome			
M26 or	*M26*	73	5300
M375 +	*M375*	84	590
+ *M210*			
Chromosome × multicopy plasmid			
M26 or	*M26*	350	14,000
M375 +	*M375*	350	1000
+ *LA69*			

[a]Standard crosses were analyzed for Ade$^+$ spores by differential plating. In the lower pair of crosses *ade6-LA69* was present on a multicopy plasmid in homothallic strains. The data are from 19.

interacts with one or more other sites. These experiments used 3.2-kilobase (kb) DNA fragments containing the entire *ade6* gene bearing one or another mutant allele and about 500 bp flanking each side of *ade6*. In one case the fragments were moved to a multicopy plasmid; recombination of this plasmid-borne *ade6* gene with the endogenous chromosomal *ade6* gene is stimulated when *M26* is on the *chromosome* but not when it is on the *plasmid* (Tables III and IV). In another case, the 3.2-kb *ade6* DNA fragments were inserted into the *ura4* gene, located about 200 cM to the left of *ade6* on chromosome III; strains containing heteroallelic *ade6* insertions were crossed to test *M26* hot spot activity at the *ura4* locus. A large deletion of the endogenous *ade6* gene prevented recombination between the inserted DNA and the endogenous *ade6* locus. This arrangement therefore provided homology across the entire recombining chromosomes. When located at *ura4*, *M26* did not have hot spot activity (i.e., it did not yield more recombinants than did *M375*) (Table IV). [The modestly (twofold) higher recombinant frequency with *M375* than with *M26* may reflect differences in mismatch correction efficiency of these alleles.] These results, plus the fact that the *M26* mutation is 1010 bp from the nearer end of the 3.2-kb fragment (18), show that at least one nucleotide located at least 1.0 kb away from *M26* is essential for the action of *M26*.

Two explanations for the context dependence of *M26* action come to mind: (1) *M26* may be part of a two-element system; i.e., *M26* may interact with another site. In this view, the second site is correctly positioned near *ade6* but not on the multicopy plasmid or near *ura4*. For *M26* to be active, the two

TABLE IV
Action of *M26* Located at *ade6* or at *ura4*[a]

Cross	Test allele	Ade$^+$ per 10^6 viable spores
Chromosome × chromosome at *ade6*		
M26 or *M375* ——— +	*M26*	10,600
+ ——— *L469*	*M375*	680
Chromosome × chromosome at *ura4*		
M26 or *M375* ——— + (inserted in *ura4*; del *ade6*)	*M26*	1900
+ ——— *L469* (inserted in *ura4*; del *ade6*)	*M375*	3500
Chromosome × multicopy plasmid (*ade6*)		
(*M26* or *M375* ——— +)	*M26*	310
+ ——— *L469*	*M375*	360

[a]Standard crosses were analyzed for Ade$^+$ spores by differential plating. In the middle pair of crosses a 3.2-kb DNA fragment containing the mutant *ade6* gene was inserted into the *ura4* locus, and the endogenous *ade6* locus was deleted. In the bottom pair of crosses *ade6-M26* or *M375* was present on a multicopy plasmid in homothallic strains. The data are summarized from 22.

sites may need to be correctly oriented with respect to each other or have the proper distance between them. Precedence for such an orientation-dependent, two-element system regulating homologous recombination is provided by the Chi and *cos* sites of *E. coli* and its phage λ. *cos* acts as an oriented entry site for RecBCD enzyme, which travels unidirectionally along the DNA and cuts one strand when encountering a properly oriented Chi site; the cut strand, unwound by RecBCD enzyme, is hypothesized to synapse with the homolog and, with further processing, to generate recombinants (5). The genetic properties of *M26* are similar to those of Chi, and the hypothetical *M26*-activating site may be an enzyme-entry site similar to *cos*.

(2) The activity of *M26* may depend in a complicated way on its chromosomal context. For example, the chromatin structure (binding of histones and other proteins, local supercoiling density, etc.) may be determined by many parts of the nucleotide sequence, and the chromatin structure may in

turn determine the action of *M26*. In this view, the bulk of the DNA surrounding *M26*, not just a single discrete site, is essential for the action of *M26*. Such context dependence of *M26* action may be similar to that of eukaryotic transcription, which can be dependent on the location of the tested gene in the genome (e.g., see 23).

Distinguishing these two explanations for the position-dependent action of *M26* may be possible by moving *M26* to other sites or by moving other chromosomal fragments near an inactive *M26* site. Identifying one or more fragments activating *M26* would allow a search for a single, common element activating *M26*.

SUMMARY

We have reviewed here our initial studies of the molecular basis of meiotic recombination in *S. pombe*. By isolating meiotic recombination-deficient (Rec^-) mutants, we have identified genes whose products are required for meiotic recombination. Enzymatic activities that have plausible roles in recombination and that are induced or activated during meiosis have been identified. We have further characterized a site (*M26*) locally stimulating meiotic recombination, and we have inferred a second site (or set of sites) required for the action of *M26*. Further studies of the genes, enzymes, and sites involved in meiotic recombination will elucidate the mechanism of meiotic recombination and its control.

ACKNOWLEDGMENTS

We are grateful to Jürg Kohli (University of Bern) for generously sharing *S. pombe* strains, plasmids, and unpublished information. We thank Nancy Hollingsworth, Edgar Käslin (University of Bern), Jürg Kohli, Karen Larson, Takehiko Shibata (Riken Institute), Rolf Sternglanz (SUNY at Stony Brook), Philippe Szankasi, Jeff Virgin, and Wayne Wahls for allowing us to cite their unpublished information and Sue Amundsen, Karen Larson, and Philippe Szankasi for helpful comments on the manuscript. Our research is supported by Grants GM31693 and GM32194 from the National Institutes of Health.

REFERENCES

1. Kohli, J. (1987). *Curr. Genet.* **11**, 575–589.
2. Leupold, U. (1970). *In* "Methods in Cell Physiology" (D. M. Prescott, ed.), Vol. 4, pp. 169–177. Academic Press, New York.
3. Iino, Y., and Yamamoto, M. (1985). *Mol. Gen. Genet.* **198**, 416–421.

4. Gutz, H. (1971). *Genetics* **69,** 317–337.
5. Smith, G. R. (1988). *In* "The Recombination of Genetic Material" (K. B. Low, ed.), pp. 115–154. Academic Press, New York.
6. Grossenbacher-Grunder, A. M., and Thuriaux, P. (1981). *Mutat. Res.* **81,** 37–48.
7. Gutz, H., and Schmidt, H. (1985). *Curr. Genet.* **9,** 325–331.
8. Ponticelli, A. S., and Smith, G. R. (1989). *Genetics* **123,** 45–54.
9. DeVeaux, L. C., Hoagland, N. A., and Smith, G. R. (1992). *Genetics,* **130,** 251–262.
10. Gutz, H., Heslot, H., Leupold, U., and Loprieno, N. (1974). *In* "Handbook of Genetics" (R. C. King, ed.), Vol. 1, pp. 395–446. Plenum Press, New York.
11. Niwa, O., and Yanagida, M. (1985). *Curr. Genet.* **9,** 463–470.
12. Catcheside, D. G. (1977). "The Genetics of Recombination." University Park Press, Baltimore, Maryland.
13. Smith, G. R. (1988). *Microbiol. Rev.* **52,** 1–28.
14. Kucherlapati, R., and Moore, P. D. (1988). *In* "Genetic Recombination" (R. Kucherlapati and G. R. Smith, eds.), pp. 575–595. American Society for Microbiology, Washington, D.C.
15. Szankasi, P. and Smith, G. R. (1992). *J. Biol. Chem.*, **267,** 3014–3023.
16. Cao, L., Alani, E., and Kleckner, N. (1990). *Cell* **61,** 1089–1101.
17. Sun, H., Treco, D. and Szostak, J. W. (1991). *Cell* **54,** 1155–1161.
18. Szankasi, P., Heyer, W.-D., Schuchert, P., and Kohli, J. (1988). *J. Mol. Biol.* **204,** 917–925.
19. Ponticelli, A. S., Sena, E. P., and Smith, G. R. (1988). *Genetics* **119,** 491–497.
20. Schuchert, P., Langsford, M., Käslin, E., and Kohli, J. (1991). *EMBO J.* **10,** 2157–2163.
21. Schuchert, P., and Kohli, J. (1988). *Genetics* **119,** 429–436.
22. Ponticelli, A. S., and Smith G. R. (1992). *Proc. Natl. Acad. Sci. U.S.A.* **89,** 227–231.
23. Laurie-Ahlberg, C. C., and Stam, L. F. (1987). *Genetics* **115,** 129–140.

15

Genetics and Molecular Biology of Recombination

ROBIN HOLLIDAY[1]
C.S.I.R.O. Laboratory for Molecular Biology
Division of Biomolecular Engineering
North Ryde, NSW 2113, Australia

INTRODUCTION

There are at least three reasons that recombination is of interest to geneticists. First, they have traditionally exploited recombination as a tool to map genes along the length of the chromosome, and this can be done without any knowledge of the actual mechanism. The same applies to recent studies of targeted recombination in mammalian cells, in which the aim is to integrate DNA at a homologous region, and the experimentalist need not be concerned about the molecular details of the process. Second, it is important to understand the biological role of recombination, particularly at meiosis. This is a problem for the evolutionary biologist and raises a number of controversial issues. Is recombination primarily a repair mechanism? Is it essential for producing genetic diversity? Or is it merely a mechanical device for ensuring the accurate disjunction of chromosomes of the first meiotic division? The third approach to recombination is the study of molecular mechanism, and this area will be briefly reviewed here.

Historically, there have been four main stages in studies of mechanisms of homologous recombination. (1) Much important information has been gained by genetic analysis itself, particularly recombination in short regions of the chromosome and the recovery of all four products of single meiotic divisions. (2) Mutants have been isolated that are altered, defective, or deficient in the process of recombination. (3) This was followed later by the

1. Based on the concluding address given at the P&S Biomedical Sciences Symposium, "Mechanisms of Eukaryotic DNA Recombination."

identification of proteins or enzymes necessary for recombination, and their inactivity in defined mutants. (4) Finally, many molecular techniques are now being employed to examine in more detail various aspects of the recombination process, including, of course, the testing of particular models of recombination. I shall briefly cover each of these approaches, illustrating stages 2–4 from experimental studies with the basidiomycete smut fungus *Ustilago maydis*.

RECOMBINATION AND HETERODUPLEX DNA

During the period when recombination was used as a tool for mapping genes on chromosomes, it was generally thought that crossing-over occurred between the genes and not within them. This completely changed with the elucidation of the structure of DNA by Watson and Crick in 1953. It was then apparent that mutations in single genes should be present in a one-dimensional array. This was realized by Pontecorvo [reviewed in (1)], and with his associates, he used the ascomycyte *Aspergillus nidulans* to identify noncomplementing mutant alleles with a common phenotype. With auxotrophic mutants, selective procedures could be used to recover rare wild-type recombinants from crosses between different mutants. Pritchard (2) was particularly successful in these studies, and he constructed a convincing one-dimensional fine structure map of the *ad8* locus. This demonstrated the equivalence of the DNA structure and the genetic map. Benzer (3,4) used the rII region of bacteriophage T4 in a much more detailed study, in which a very large number of mutants were mapped in a linear array.

Throughout the 1950s and early 1960s, a number of experimental studies were carried out with fungi in which tetrads could be analyzed. These included *Neurospora*, *Sordaria*, *Ascobolus*, and a little later on, *Saccharomyces*. These experiments showed that there was a reciprocal mode of recombination (crossing-over) and a nonreciprocal mode (gene conversion), which were closely related processes, or different outcomes of the same process, involving two of the four meiotic chromatids at any one exchange point. Crossing-over within a gene would of course be associated with the exchange of outside markers, and gene conversion is frequently associated with such exchange (approximately 50%). The studies convincingly demonstrated the occurrence of postmeiotic segregation, where sister spores derived from a single meiotic product were mutant and wild type, suggesting that they arose from a single heteroduplex DNA structure, equivalent to a bacteriophage heterozygote. Also, it was shown that gene conversion was polarized, in that the frequency of events at one end of the gene was often higher than that at the other end.

It was fashionable during this period to explain gene conversion with associated crossing-over by a copy choice mechanism, in which DNA synthesis was presumed to be conservative (1, 5–7). It was proposed that newly synthesized molecules could switch from one parental template to the other with occasional miscopying to give three parental genotypes, rather than two, in a tetrad. It was curious that these models were taken seriously, although it was already known at the time that DNA replication was semiconservative. Moreover, there was nothing in the copy choice models to indicate how the switching could be accurate, that is, without the gain or loss of genetic information. Innumerable studies had established beyond doubt that recombination was a precise process at the molecular level, which did not generate duplications or deletions of genetic material.

From a consideration of real DNA structures it was evident that the process of molecular synapses was likely to involve complementary base pairing. A mechanism was needed in which single strands come apart and annealed with opposite partners. The points where polynucleotide strands exchange pairing partners could provide the position for the exact breakage and reunion of strands, with a 50% probability of outside marker exchange. If the heteroduplex or hybrid DNA included parental mutant and wild-type alleles, then mismatched bases would be formed. These might segregate to give postmeiotic segregation, or they might be repaired to produce gene conversion. An early model of this type consisted of a structure with no free ends: it was presumed that Watson–Crick strands unraveled and paired with an opposite partner to produce a symmetrical "bubble," with regions of hybrid DNA of equal length flanked by two polynucleotide exchange points (Fig. 1). This was not published, because there was no obvious way in which such a structure could arise from two molecules consisting only of right-hand helices. At that time topoisomerases were unknown, as was left-handed Z DNA (see the following). The later published model assumed that single-strand nicks occurred at homologous defined sites, which allowed unraveling and reannealing of single polynucleotide chains to form again symmetrical regions of heteroduplex DNA, but with only one exchange point (14). Such a model could account for polarized gene conversion if the hybrid DNA originating at the initial break points was of variable length. It should be emphasized that other more recent models of recombination vary primarily in the mode of initiation of the recombination structure. Thus, the Meselson and Radding (15) model depends on one single-strand break to produce non-reciprocal hybrid DNA, which later can generate a symmetrical structure, and the model of Szostak *et al.* (10) depends on an initial double-strand break in one parental molecule.

The exact structure of the exchange point has recently become controversial. Sigal and Alberts (16) and Jeggo and Banks (17) built molecular models

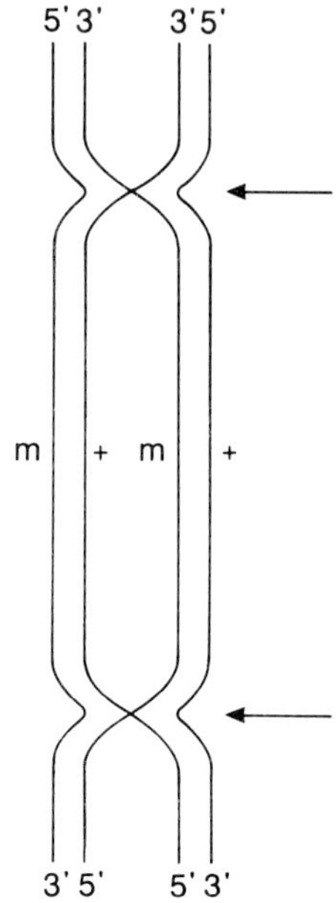

Fig. 1. A recombination intermediate, first proposed in 1959, involving the formation of reciprocal regions of hybrid or heteroduplex DNA. If such regions include a heterozygous marker, then mismatched bases will occur. Such mismatches might be repaired, to produce gene conversion, or separate at the next round of DNA synthesis to produce postmeiotic segregation, including aberrant 4:4s. The arrows indicate points where polynucleotide chains of like polarity could break and rejoin to produce crossing-over or non-crossing-over of outside markers with equal frequency. The structure has been proposed as a recombination intermediate in various contexts. For instance, it can be generated following breakage and reunion of polynucleotide chains (8–10). It could be produced by topoisomerase activity (11), or it could be the result of synapsis to form a paranemic joint, with both right- and left-handed (Z DNA) helical regions (12), which can subsequently be stabilized by topoisomerase to form a plectonemic joint (13).

that showed that polynucleotide chains could exchange pairing partners while retaining complete hydrogen bonding (Fig. 2). The exchange could diffuse in either direction, with rotation of the arms, by a process now known as branch migration (18,19). Sobell (20) clarified the process of isomerization, showing that it involves a 180° rotation of arms to form the cruciform intermediate, and a further 180° rotation to generate the isomer, in which the strands previously on the outside of the exchange point are now on the inside (see Fig. 2). These models have been widely accepted, because the cruciform structure was visualized by electron microscopy by Potter and Dressler (21), as well as several other investigators, and also the process of branch

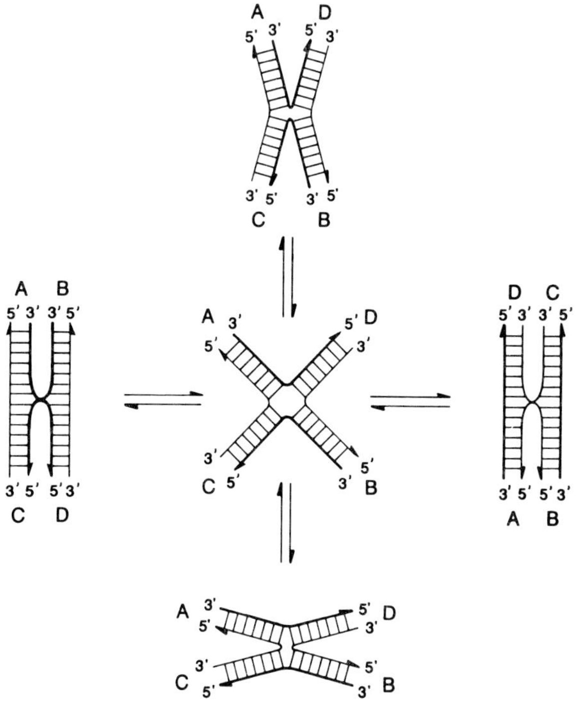

Fig. 2. The two isomeric forms of the Sigal and Alberts structure (16) are shown left and right. Sobell (20) demonstrated that the isomerization reaction requires 180° rotation of arms to form a cruciform intermediate, followed by a further 180° rotation. When the double helices are homologous, all these structures can branch migrate by the breaking and reforming of hydrogen bonds (18,19). It has been proposed that immobile junctions that cannot branch migrate (22) may form the structures shown at the top and bottom (23). Isomerization of these involves simple opening or closing of arms. It is unlikely that these structures are formed during recombination between homologous DNA molecules, because they are essentially static configurations, which lack the mobility required for branch migration of recombination intermediates (see text).

migration had been demonstrated. Another way of studying the structure is to construct oligonucleotides, which are not homologous, but can form immobile junctions, in which branch migration is no longer possible. Complementary base pairing occurs in all four arms, but movement of the exchange point is impossible because of the lack of homology in opposite arms of the cruciform structure (22). Duckett *et al.* (23) included different restriction enzyme sequences in each arm so that they could be specifically cut and shortened at will. On the basis of the theory of migration of DNA structures in agarose gels, which is certainly a matter for debate and may well be ill-founded (H. Drew, personal communication), they concluded that the immobile junction formed structures shown at the top and bottom of Fig. 2. These have a severe kink in the polynucleotide chain, and molecular models have not been built to show whether this is compatible with complete base pair stacking, which is an essential feature of their proposal. Furthermore, in a real recombination intermediate between homologous DNA molecules, it is difficult to see how branch migration could occur in the structure proposed. The individual bases would have to traverse the kinks in the backbone through 180° rotation, and there would be severe disruption of base stacking during this movement. On the other hand, it is easy to see how hydrogen bonds between complementary base pairs could smoothly melt and reform in the Sigal and Alberts structure (16) without any change in the plane of the bases and much better maintenance of the base stacking during branch migration.

MUTANTS DEFECTIVE IN RECOMBINATION

It became clear in the early 1960s that the formation and resolution of recombination intermediates must depend on enzymes or other proteins. In *Escherichia coli* Clark and Margulies (24) isolated the first recombination-deficient mutants. There was no direct method to obtain such mutants in fungi, but as there were clear affinities between repair and recombination, I decided to isolate repair-deficient mutants of *U. maydis*, using a simple replica plating procedure (25), and these were the first to be obtained in any eukaryotic organism. Two of these mutants, originally designated *uvs-1* and *uvs-2*, were soon found to be defective in mitotic recombination (26), and they were subsequently renamed *rec1* and *rec2*. Both were recessive, but when homozygous in diploids, each blocked radiation-induced allelic recombination. Diploids homozygous for both *rec1* and *rec2* had a fully rec^- phenotype, in that they were incapable of spontaneous allelic recombination as well (26,27). Crosses between *rec2* strains were totally blocked in meiosis, and those between *rec1* strains produced many aneuploids and nonviable products. The phenotype of *rec1* was extensively characterized, and it was shown to have pleiotropic effects on the genetic system of the organism (28). These results will not be reviewed here, but one important feature was the finding that the spontaneous mitotic recombination was defective and frequently associated with chromosome or chromatid breakage, presumably of one of the DNA molecules involved in the exchange. Thus, the precision of the recombination seen in wild-type strains was lost. Also *rec1* produced striking instability in diploid strains, which is probably because of chromosome breaks during normal growth, together with the generation of aneuploids. These and other features of the phenotype of *rec1* led to the conclusion that the wild-type gene had an essential regulatory role as well as being necessary for normal recombination (28). In this regard there are similarities to the *recA* mutants of *Escherichia coli*.

RECOMBINATION PROTEINS

After the isolation and characterization of *rec*$^-$ mutants of *E. coli*, there was a long period before the proteins produced by the wild-type genes were identified. Subsequently the recA protein was shown to be essential for the synapsis of homologous DNA molecules, and it is also essential for the SOS repair response in wild-type cells [reviewed in 29]. The first eukaryotic protein with related properties was isolated from *U. maydis* (30). It carries out a series of ATP-dependent annealing or strand-transfer reactions involving homologous DNA molecules (13, 31–34). It was also shown that molecules with internal homology, but nonhomologous ends, can synapse in the presence of *rec1* protein and adenosine triphosphate (ATP). These reactions suggested that the synapsed molecules had a paranemic rather than a plectonemic structure. Such an intermediate could be produced by equal stretches of B and Z DNA (right- and left-handed helices), and evidence that Z DNA might be involved in the reaction was obtained using antibody to this structure (13, 33, 34). Also, *rec1* protein binds more tightly to Z DNA than to B DNA, suggesting that the former is an intermediate in the synapsis reaction. Furthermore, supercoiled plasmids, containing sequences of alternating purine and pyrimidine bases that can form Z DNA, were shown to form synapsed complexes. The addition of topoisomerase to the reaction resulted in the formation of linked circles, which were stable after cross-linking the DNA with psoralen and cutting the circles at a single restriction enzyme site. The results suggested that the recombination structure produced by *rec1* protein and topoisomerase was that shown in Fig. 1 (13, 32). It is interesting that this reciprocal recombination structure, which was initially considered 30 years ago, has for rather different reasons often been proposed as a recombination intermediate (8–13).

The *rec1* protein takes its name from an early demonstration that *rec1* strains lacked activity (30). Reversion of the first two alleles *rec1-1* and *rec1-2* were not seen, but another allele, *rec1-5*, did produce a radiation-resistant derivative. This proved to be due to an unlinked suppressor, which has not been further categorized, but it was shown that the *rec1-5* suppressed strain had regained *rec1* activity, using three different assay procedures (27).

MOLECULAR ANALYSIS

The molecular genetics of *U. maydis* has been developed in three laboratories and an efficient transformation system is now available (35–41). These procedures can be used to study further the *rec1* gene and determine whether it codes for the *rec1* protein. Both the *rec1* and the *rec2* wild-type genes have been cloned by a direct complementation procedure. Disruption of the chromosome wild-type *rec1* gene has also been achieved, using a hygromycin-resistance gene insert, and this disrupted gene does not complement *rec1* mutants, thus proving that the cloned gene is indeed *rec1*$^+$ (41). Also, a *rec1* mutant transformed by the wild-type gene has regained all the normal wild-type phenotypic characteristics. The transcript in wild-type strains is 1.7 kilobases (kb), and the level of transcription increases severalfold 4–6 hr after treatment of cells with UV light. (Recombination in heteroallelic diploids increases dramatically after UV treatment, but not in isogenic diploids homozygous for *rec1* (28).) The transcript size is somewhat shorter than that expected for the *rec1* protein, which has a molecular weight of 70,000. Subsequent studies have demonstrated that the *rec1* gene does not code for the *rec1* protein and the latter will be renamed.

Analysis of transformation of *leu*$^-$ strains of *U. maydis* using a *leul*$^+$ gene in a plasmid without an ARS sequence, has revealed that there are three pathways, namely, homologous integration, nonhomologous integration, and also the formation of concatamers, which appear to replicate autonomously (42). The formation of such concatamers is of particular interest, since it is well known that mammalian cells can produce such structures by the joining of nonhomologous ends of DNA. Further studies on the pathways of DNA transformation in *U. maydis* are in progress using the *leu*$^+$ selection system and recipient cells that are either *rec1* or *rec2*. It is gratifying that these mutants, isolated over 25 years ago, will now be invaluable in the further analysis of molecular pathways of DNA transformation. This information may well make it easier to understand the mechanisms of homologous and nonhomologous integration of exogenous DNA into chromosome of mammalian cells.

ACKNOWLEDGMENTS

I thank William Holloman, Scott Fotheringham, Toyoko Tsukuda, and Robert Banchwitz for helpful discussions and for bringing me up to date with their recent molecular studies of recombination in *Ustilago maydis*. I also thank Horace Drew and Richard Cowan for discussions on the likely effects of DNA structure on migration in gels.

REFERENCES

1. Pontecorvo, G. (1958). "Trends in Genetic Analysis." Columbia University Press, New York.
2. Pritchard, R. H. (1955). *Heredity* **9,** 343–371.
3. Benzer, S. (1955). *Proc. Natl. Acad. Sci. U.S.A.* **41,** 344–354.
4. Benzer, S. (1961). *Proc. Natl. Acad. Sci. U.S.A.* **47,** 403–515.
5. Freese, E. (1957). *Genetics* **42,** 671–684.
6. de Serres, F. J. (1958). *Cold Spring Harbor Symp. Quant. Biol.* **23,** 111–118.
7. Pritchard, R. H. (1960). *Symp. Soc. Gen. Microbiol.* **10,** 155–180.
8. Sobell, H. M. (1972). *Proc. Natl. Acad. Sci. U.S.A.* **69,** 2483–2487.
9. Holliday, R. (1974). *Genetics* **78,** 273–287.
10. Szostak, J. W., Orr-Weaver, T. L., Rothstein, R. B., and Stahl, F. W. (1983). *Cell* **33,** 25–35.
11. Champoux, J. J. (1977). *Proc. Natl. Acad. Sci. U.S.A.* **74,** 5329–5332.
12. Pohl, F. M. (1969). *Naturwissenschaften* **54,** 616.
13. Kmiec, E. B., and Holloman, W. K. (1986). *Cell* **44,** 545–554.
14. Holliday, R. (1964). *Genet. Res. Camb.* **5,** 282–304.
15. Meselson, M. J., and Radding, C. M. (1975). *Proc. Natl. Acad. Sci. U.S.A.* **72,** 359–361.
16. Sigal, N., and Alberts, B. (1972). *J. Mol. Biol.* **71,** 789–793.
17. Jeggo, P. A., and Banks, G. R. (1970). Unpublished observations.
18. Meselson, M. (1972). *J. Mol. Biol.* **71,** 795–798.
19. Thompson, B. J., Carmien, M. N., and Warner, R. C. (1976). *Proc. Natl. Acad. Sci. U.S.A.* **73,** 2299–2303.
20. Sobell, H. M. (1974). *In* "Mechanisms in Recombination" (R. F. Grell, ed.), pp. 433–438. Plenum Press, New York.
21. Potter, H., and Dressler, D. (1976). *Proc. Natl. Acad. Sci. U.S.A.* **73,** 3000–3004.
22. Kallenbach, N. R., Rong-Ine, M., and Seeman, N. C. (1983). *Nature (London)* **305,** 829–831.
23. Duckett, D. R., Murchie, A. I. H., Diekmann, S., von Kitzing, E., Kemper, B., and Lilley, D. M. J. (1988). *Cell* **55,** 79–89.
24. Clark, A. J., and Margulies, A. D. (1965). *Proc. Natl. Acad. Sci. U.S.A.* **53,** 451–459.
25. Holliday, R. (1965). *Mutat. Res.* **2,** 557–559.
26. Holliday, R. (1967). *Mutat. Res.* **4,** 275–288.
27. Holliday, R., Taylor, S. Y., Kmiec, E. B., and Holloman, W. K. (1984). *Cold Spring Harbor Symp. Quant. Biol.* **49,** 669–673.
28. Holliday, R., Halliwell, R. E., Evans, M. W., and Rowell, V. (1976). *Genet. Res.* **27,** 413–453.
29. Radding, C. M. (1982). *Annu. Rev. Genet.* **16,** 405–437.
30. Kmiec, E. B., and Holloman, W. K. (1982). *Cell* **29,** 367–374.
31. Kmiec, E. B., and Holloman, W. K. (1983). *Cell* **33,** 857–864.
32. Kmiec, E. B., Kroeger, P. E., Broughman, M. J., and Holloman, W. K. (1983). *Cell* **34,** 919–929.
33. Kmiec, E. B., and Holloman, W. K. (1984). *Cell* **36,** 593–598.
34. Kmiec, E. B., Angelides, K. J., and Holloman, W. K. (1985). *Cell* **40,** 139–145.
35. Banks, G. R. (1983). *Curr. Genet.* **7,** 73–77.
36. Banks, G. R., and Taylor, S. Y. (1988). *Mol. Cell. Biol.* **8,** 5417–5424.
37. Wang, J., Holden, D., and Leong, S. A. (1988). *Proc. Natl. Acad. Sci. U.S.A.* **85,** 865–869.

38. Holden, D. W., Kronstad, J. W., and Leong, S. A. (1989). *EMBO J* **8,** 1927–1934.
39. Tsukuda, T., Carleton, S., Fotheringham, S., and Holloman, W. K. (1988). *Mol. Cell. Biol.* **8,** 3703–3709.
40. Fotheringham, S., and Holloman, W. K. (1989). *Mol. Cell. Biol.* **9,** 4052–4055.
41. Tsukuda, T., Bauchwitz, R., and Holloman, W. K. (1989). *Gene* **85,** 335–341.
42. Fotheringham, S., and Holloman, W. K. (1989). *Genetics* **124,** 833–843.

PART V

GENETIC RECOMBINATION IN TRYPANOSOMES AND *PLASMODIUM*

16

Chromosome and Telomere Structure in *Trypanosoma brucei*

LEX H. T. VAN DER PLOEG[1], KEITH GOTTESDIENER[1], DORIS B. TSE[2], HUI-MIN CHUNG[1], AND MICHAEL WEIDEN[1]

1. *Department of Genetics and Development and Department of Medicine Columbia University New York, New York*

2. *Laboratory of Lymphocyte Cell Biology Department of Medicine North Shore University Hospital Cornell University Medical College, Manhasset, New York 11030*

INTRODUCTION

African trypanosomes are protozoan flagellates that proliferate in the blood of mammalian hosts. These protozoan parasites escape immunodestruction by antigenic variation of their major cell surface protein, the variant cell surface glycoprotein or VSG (for reviews see Parsons *et al.*, 1984; Boothroyd, 1985; Donelson and Rice-Ficht, 1985; Borst and Greaves, 1987; Clayton, 1987; Pays and Steinert, 1988; Cross, 1990; Van der Ploeg, 1990).

We have studied chromosome and nuclear structure in trypanosomes, which will aid the genetic analysis of these organisms and provide insights into mechanisms of chromosome (in)stability and maintenance in protozoa (Van der Ploeg *et al.*, 1989; Gottesdiener *et al.*, 1990; Chung *et al.*, 1990). Given the development of methods for the transient (Bellofatto and Cross, 1989; Laban and Wirth, 1989; Kapler *et al.*, 1990; Rudenko *et al.*, 1990; Clayton *et al.*, 1990) and stable transfection by homologous recombination (ten Asbroek *et al.*, 1990; Lee and Van der Ploeg, 1990; Eid and Sollner-Webb, 1991; Cruz and Beverley *et al.*, 1990; Tobin *et al.*, 1991) in several different trypanosomatid species, a detailed analysis of protozoal genetics and trypanosome biology is now possible. In this chapter we present an overview of chromosome content and structure in the parasitic protozoan *Trypanosoma brucei*.

ORGANIZATION OF THE *TRYPANOSOMA BRUCEI* GENOME; PRESENCE OF CHROMOSOME HOMOLOGS

The genome of both the bloodstream form and insect form of the protozoan *Trypanosoma brucei* (stock 427–60) is diploid and consists of approximately 120 mitotically stable chromosomes (Borst *et al.*, 1982; Van der Ploeg *et al.*, 1984a, 1989). One hundred of these chromosomes are small, 50–150 kilobase (kb) minichromosomes (Williams *et al.*, 1982; Van der Ploeg, 1984a), for which the genome may be considered aneuploid (Gottesdiener *et al.*, 1990). Most of the roughly 20 larger chromosomes (200 kb up to at least 5.7 Mb), accounting for 90% of the genome, can be grouped into homologous pairs (Gottesdiener *et al.*, 1990). The identification of chromosome homolog was based on three criteria: (1) several housekeeping genes could each be shown to distribute to two differently sized chromosomes, generating specific linkage groups (Table I); (2) for most of these genes the physical maps of the flanking regions could be shown to be virtually identical over hundreds of kilobasepairs on the two different chromosomes (Gottesdiener *et al.*, 1990); (3) for one of the largest chromosome bands (band 19, at least 5.7 Mb), which could not be further separated by pulsed field gradient (PFG) electrophoresis, *in situ* hybridization revealed two distinct sites of hybridization in the interphase nucleus, indicative of alleles of genes on homologous chromosomes. For instance, the tandem arrays of heat shock 70 (*hsp70*) protein-coding genes (Glass *et al.*, 1986), mapped to band 19 alone in the PFG electrophoresis. On the basis of the nonstoichiometric ethidium bromide staining intensity, PFG band 19 was proposed to consist of more than a single chromosome. The homologs encoding the *hsp70* alleles could therefore comigrate at band 19. To confirm this hypothesis we performed *in situ* hybridization with digoxigenylated DNA probes and applied indirect immunofluorescence to identify the chromosomal location of the *hsp70* genes in the interphase *T. brucei* nucleus. As expected for alleles of *hsp70* genes on homologous chromosomes, the hybridization revealed two distinct signals in the majority of interphase (G_1 and S phase) nuclei (Fig. 1; see dots marked with arrowheads in several nuclei, in the right panel; see figure legend for details). These data, together with the chromosome assignments shown in

TABLE I
Distribution of Genetic Markers on Different Chromosomes of *T. brucei* Stock 427-60[a]

Band	Gene	Chromosome size (Mb)
19′*	TIM, HSP70, rRNA	5.7
19*	ALDO, CRAM, cGAPDH, PARP B, PYK	
18	ALDO, CRAM, cGAPDH, PARP B, PYK	
17	medRNA	
16′*	PGI, TUB, TBS18	
16*	TBS17, rRNA	
15′*	TBS17, rRNA	3.0
15*	gGAPDH, PARP A	
14′*	medRNA	
14*	gGAPDE, PARP A	
13		
12		
11	PGI, TUB, TBS18	1.5
10*	ES 118, rRNA	
9′*	ES 118, rRNA	
9*		
8	TBS4, PLC	
7	TBS4, PLC	
6		0.5
5		
4		
3		
2		
1	(MINICHROMOSOMES)	0.15

[a]Sizes of chromosomes are in megabase pairs (Mb). Asterisks identify nonstoichiometrically, ethidium bromide-straining PFG bands. Chromosomes in band 14, labeled 14 and 14′, have been shown to size separate in some variants (Gottesdiener *et al.*, 1990; chromosome 9 and 9′ can be separated in all variants tested, depending on the PFG conditions chosen; in all other cases "apostrophe" refers to the presumed second chromosome of a nonstoichiometrically staining band. Determination of the chromosomal location of the rRNA genes was performed with a probe encoding an entire rRNA repeat unit. Only the strong hybridizing bands have been listed. Several rRNA hybridizing bands light up in longer exposures, which we assume encode rRNA orphons. Markers that uniquely hybridize with band 19 are underlined. Evidence for the presence of the arrays of *Hsp70* genes on different chromosomes has been presented in Fig. 1. Chromosome pairs are indicated with brackets, showing the markers for each pair. Probes used: ALDO, a gene encoding fructose-bisphosphate aldolase; cGAPDH, a gene encoding cytosolic glyceraldehyde-phosphate dehydrogenase, CRAM, a cDNA encoding a putative cell surface receptor of the flagellar pocket; ES 118, a genomic fragment obtained from a band 10 specific library; gGAPDH, a gene encoding the glycosomal glyceraldehyde-phosphate dehydrogenase; *Hsp70*, heat shock 70 protein-coding gene; medRNA, a genomic miniexon donor RNA gene; PARP, probes specific for the loci from the procyclic acidic repetitive protein A and B genes; PGI, a gene encoding glucose-phosphate isomerase; PLC, the phospholipase C gene; PYK, the gene for pyruvate kinase; rRNA, a rRNA repeat gene; TBS 4, a protein-coding gene of unknown function; TBS 17, a gene encoding a protein that locates to the flagellum of *T. Brucei;* TBS 18, a cDNA encoding a protein of unknown function; TIM, triose-phosphate isomerase; TUB, an α-β-tubulin genomic clone (see Gottesdiener *et al.*, 1009 for original references describing each marker used).

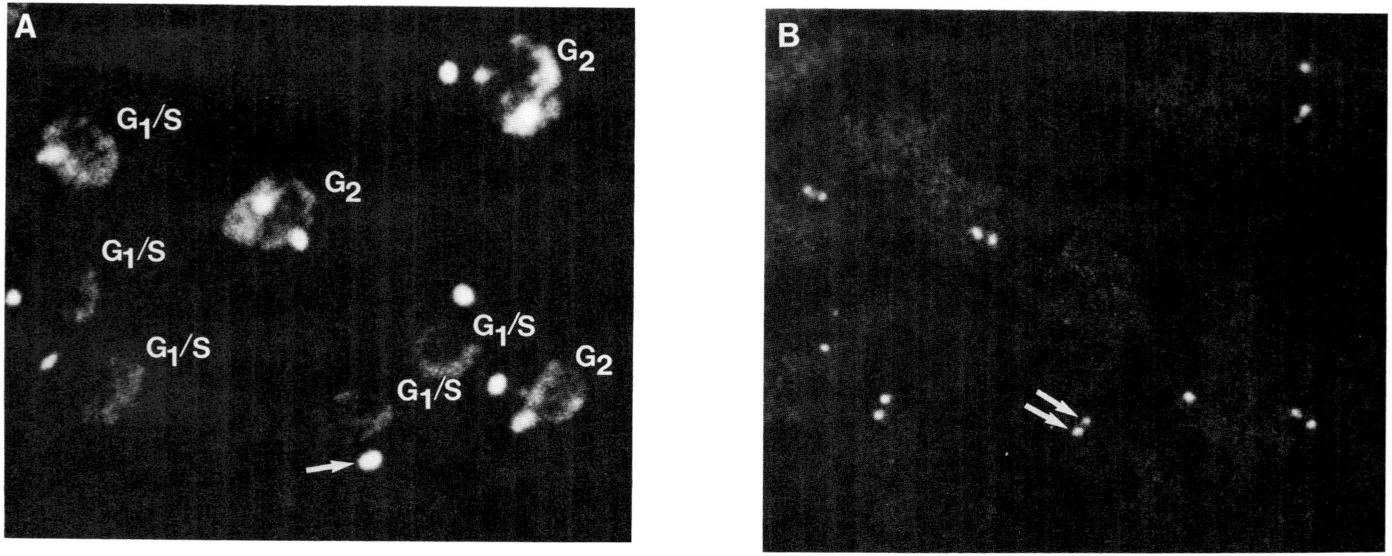

Fig. 1. Nuclear location of hsp70 genes in procyclic trypanosomes. *In situ* hybridization was performed according to Chung *et al.* (1990) with modifications. A nick-translated digoxigenin-derivatized *hsp70* probe (clone hl; Glass *et al.*, 1986) was hybridized to cytosmears of methanol; acetic acid fixed procyclic trypanosomes. The hybridization signal was detected with mouse antidigoxigenin monoclonal antibodies and fluorescein-conjugated goat antimouse immunoglobulin. (A) 4′, 6′-Diamidino-2-phenylindole (DAPI) staining of nuclei and kinetoplasts in procyclic trypanosomes. The phase of the cell cycle of each trypanosome was determined by comparing the number of kinetoplasts and nuclei in each cell (G_1/S phase, one nucleus and one kinetoplast; G_2 phase, one nucleus, two kinetoplasts; M phase, two nuclei and two kinetoplasts; as marked at the top right side of each nucleus). Borst *et al.*, 1980, 1982; Chung *et al.*, 1990). The arrow highlights a kinetoplast of a G_1/S-phase nucleus. G_2-phase cells have one nucleus and two kinetoplasts. (B) Fluorescein hybridization signals obtained with the *hsp70* probe. The arrows highlight an example of an interphase trypanosome nucleus with two distinct *hsp70* hybridization signals. More than 90% of the interphase nuclei showed two hybridization signals. Damage to nuclei, as a result of the hybridization procedure, or superimposition of hybridization signals, can lead to loss of hybridization signal. The diameter of the DAPI-stained interphase nucleus is about 2 μm.

Table I, indicate that the majority of the chromosomes, excluding the minichromosomes, can be grouped into homologous pairs.

Genetic exchange has been documented in African trypanosomes, but the actual mechanisms involved have not been elucidated (for reviews, see Le Page *et al.*, 1988; Sternberg and Tait, 1990). The presence of chromosome homolog, however, suggests the existence of gametes in genetic exchange.

Surprisingly, these chromosome homologs differed in size by at least 20% (Table I), and for several of the chromosome pairs, the size difference was as much as 50% (band 16 and 11 in Table I). The origin of this size heterogeneity is unclear. It could be restricted to the telomerically located domains, as has been proposed for size heterogeneity of the chromosomes in the protozoan *Plasmodium falciparum* (Pologe and Ravetch, 1986; Corcoran *et al.*, 1988).

MINICHROMOSOME STRUCTURE

A large repertoire of silent or basic copy (BC) VSG can be found at the telomeres of many trypanosome chromosomes, including the minichromosomes (Van der Ploeg *et al.*, 1982, 1984a, 1989; De Lange and Borst, 1982). Minichromosomes, which account for the majority of the trypanosome telomeres, have been found only in trypanosome species that undergo antigenic variation. It has been proposed that their only function is to provide the trypanosome with a large repertoire of telomerically located VSG genes (Van der Ploeg *et al.*, 1984b).

The minichromosomes are mitotically stable (Van der Ploeg *et al.*, 1984a). However, during genetic exchange their number may vary, and they, in contrast to the larger chromosomes, do not appear to follow Mendelian inheritance (Wells *et al.*, 1987). We initiated an analysis of the structure of the trypanosome minichromosomes to start (1) to define the DNA nucleotide sequence elements that are sufficient for chromosome maintenance in trypanosomes; (2) to understand the size constraints and mitotic stability of the minichromosomes and their instability during genetic exchange; and (3) to gain insight into the origin of minichromosomes.

Presumably all trypanosome chromosomes, including the minichromosomes, contain, from a telomeric location toward a chromosome internal position, an array of $(GGGTTA)_n$ repeats (Van der Ploeg *et al.*, 1984c; Blackburn and Challoner *et al.*, 1984) followed by a $(GGGTTA)_n$ derived subtelomeric repeat, with a 29-bp periodicity; adjacent to this is a 155-bp repeat composed of 74% AT-rich DNA, separated by a 74-bp repeat with a high GC content (65% GC) (Weiden *et al.*, 1991). The GC-rich DNA encodes 11 bp of the telomere GGGTTA repeat present in inverted orientation with respect to the telomere repeats. The AT-rich spacer DNA contains bent DNA.

Additional types of subtelomeric repeats are found distributed to subclasses of chromosomes only. One, a 66-bp subtelomeric repetitive restriction enzyme fragment, detected a subclass of the minichromosomes (50–100 kb in size) and a single 300-kb chromosome. Adjacent to this 66-bp fragment is the first copy of a 177-bp repeat, which is essentially minichromosome-specific and which compromises over 90% of the length of the minichromosomes. As a result of these large arrays of simple sequence DNA, the size of the majority of the (50–150 kb) minichromosomes can be reduced by several kilobases only if they are digested with frequent cutting restriction enzymes that do not cut the 177-bp repeat (enzymes like *Hin*fI, *Hae*II, and *Hae*III). The trypanosome minichromosomes may, therefore, have a limited protein-coding capacity and consist mainly of simple sequence DNA.

A model for the structure of the minichromosomes resulted from this analysis (Fig. 2; Weiden *et al.*, 1991), and several features are worth noting. First, the organization of the trypanosome telomere and subtelomere sequence elements shares a striking similarity with that of the telomeres of yeast and humans (Chan and Tye, 1983; Horowitz and Haber, 1985; Walmsley *et al.*, 1984; Cheng *et al.*, 1989; De Lange *et al.*, 1990): the trypanosome telomere repeat is identical to the telomere repeat of humans (Moyzis *et al.*, 1988), and the distribution of families of subtelomeric repeats over specific chromosomes is reminiscent of the telomeres in yeast and humans. The length of the GGGTAA-derived subtelomere repeat (29 bp) is identical to that of the subtelomere repeats in humans, though it has a different DNA nucleotide sequence (De Lange *et al.*, 1990). It is of interest that 28 bp is the minimal size of a sequence that can support recombinational extension of telomeres in yeast (Pluta and Zakian, 1989).

Second, a subclass (50–100 kb in size) of the minichromosomes mainly (> 90% of the length of the minichromosome) consisted of a 177-bp repeat, previously identified as a satellite repeat by Sloof *et al.* (1983). Evidence for other abundant repeat elements interspersed with the 177-bp repeat was not found (Weiden *et al.*, 1991). The 177-bp repeat thus becomes a possible candidate for a trypanosome centromere.

Third, several of the minichromosome telomeres analyzed did not encode VSG genes and were devoid of any protein-coding genes. The protein-coding potential of these telomeres is therefore more limited than anticipated.

Fourth, minichromosome-derived steady-state RNA could not be identified. The only function of the minichromosomes may thus be to provide for a repertoire of silent, BC VSG genes.

Finally, since many of the minichromosomes are built of similar DNA

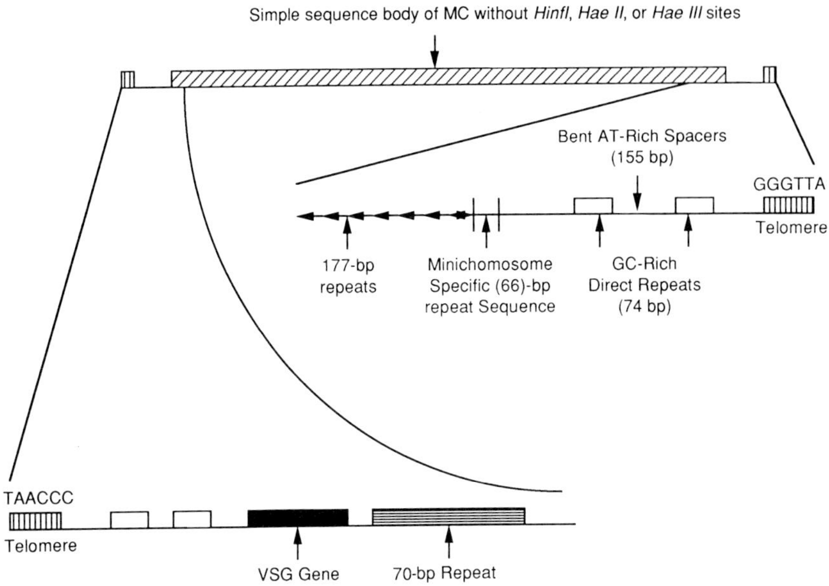

Fig. 2. Model of the structural elements in a trypanosome minichromosome (according to Weiden *et al.*, 1991). All symbols and abbreviations are explained in the figure. The presence of BC VSG genes and 70-bp repeats at some minichromosomes has been documented in Van der Ploeg *et al.* (1984a) and Van der Ploeg and Cornelissen (1984).

nucleotide sequence elements, they have been proposed to be amplified DNA, which may explain their non-Mendelian behavior during genetic exchange (Gottesdiener *et al.*, 1990; Weiden *et al.*, 1991).

An electron microscopic analysis of the minichromosomes revealed that they were linear molecules without abnormal structures in the main body of the chromosome (Weiden *et al.*, 1991). The presence of replicating molecules with a single Y or bubble structure indicated that minichromosomes may be replicated from a single centrally located origin of replication. We hypothesized that the length of a minichromosome could thus be regulated by two different parameters: first, if only a single, centrally located origin of replication is present, the upper size limit of the minichromosomes may be constrained by the time it takes to copy the entire chromosome, once per cell cycle; and second, the lower size limit may be determined by the reduced mitotic stability of very small chromosomes, as has been shown in yeast (Murray and Szostak, 1983).

Given the uniformity of DNA nucleotide sequence elements on minichromosomes, we proposed that they could represent amplified DNA. The repertoire of minichromosomes in *T. brucei* could then have resulted from the generation of a single minichromosome via the breakage and healing of a telocentric chromosome arm, followed by amplification driven by the selective advantage of having a large number of telomeric VSG genes.

CONCLUDING REMARKS

Molecular genetic analysis of parasitic protozoa has led to new insights into the organization of protozoal genomes and the understanding of chromosome structure and stability. This knowledge will facilitate the analysis of genetic crosses with *T. brucei* and lead to further understanding of the biology and molecular genetics of host–parasite interactions.

ACKNOWLEDGMENTS

We thank all colleagues from the laboratory for critical reading of the manuscript. This work was supported by NIH Grant AI 21784 and by a Grant from the John D. and Catherine T. MacArthur Foundation to LHTVDP and a Special Projects Grant from the Department of Medicine at North Shore University Hospital to DBT. LHTVDP is a Burroughs Wellcome Scholar in Molecular Parasitology.

REFERENCES

Bellofatto, V., and Cross, G. A. M. (1989). *Science* **244,** 1167.
Blackburn, E. H., and Challoner, P. B. (1984). *Cell* **36,** 447.
Boothroyd, J. C. (1985). *Annu. Rev. Microbiol.* **39,** 475.
Borst, P., and Greaves, D. R. (1987). *Science* **235,** 658.
Borst, P., Fase-Fowler, F. F., Frasch, A. C. C., Hoeijmakers, J. H. J., and Weijers, P. J. (1980). *Mol. Biochem. Parasitol.* **1,** 221.
Borst, P., Van der Ploeg, M., Van Hoek, J. F. M., Tas, J., and James, J. (1982). *Mol. Biochem. Parasitol.* **6,** 13.
Chan, C. S. M., and Tye, B-K. (1983). *Cell* **33,** 563.
Cheng, J. F., Smith, C. L., and Cantor, C. R. (1989). *Nucleic Acids Res.* **17,** 6109.
Chung, H. M., Tse, D. B., Fields, S., Taub, R. N., Shea, C., and Van der Ploeg, L. H. T. (1990). *EMBO J.* **9,** 2611.
Clayton, C. E. (1987). *Genet. Engin.* **7,** 1.
Clayton, E., Fueri, J. P., Itzhaki, J. E., Bellofatto, V., Sherman, D. R., Wisdom, G. S., Vijayasarathy, S., and Mowatt, M. R. (1990). *Mol. Cell. Biol.* **10,** 3036.
Corcoran, L. M., Forsyth, K. P., Bianco, A. E., Brown, G. V., and Kemp, D. (1988). *Cell* **53,** 807.
Cross, G. A. (1990). *Annu. Rev. Immunol.* **8,** 83–110.
Cruz, A., and Beverley, S. M. (1990). *Nature (London)* **348,** 171.
De Lange, T., and Borst, P. (1982). *Nature (London)* **299,** 451.
De Lange, T. L., Shiue, L., Myers, R. M., Cox, D. R., Naylor, S. L., Killery, A. M., and Varmus, H. E. (1990). *Mol. Cell. Biol.* **10,** 518.

Donelson, J. E., and Rice-Ficht, A. C. (1985). *Microbiol. Rev.* **49,** 107.

Eid, J., and Sollner-Webb, B. (1991). *Proc. Natl. Acad. Sci. U.S.A.* **88,** 2118–2121.

Glass, D., Polvere, R. I. P., and Van der Ploeg, L. H. T. (1986). *Mol. Cell. Biol.* **6,** 4657.

Gottesdiener, K., Garcia-Anoveros, J., Lee, G.-S. M., and Van der Ploeg, L. H. T. (1990). *Mol. Cell. Biol.* **10,** 6079.

Horowitz, H., and Haber, J. E. (1985). *Mol. Cell. Biol.* **5,** 2369.

Kapler, G. M., Coburn, C. M., and Beverley, S. M. (1990). *Mol. Cell. Biol.* **10,** 1084.

Laban, A., and Wirth, F. D. (1989). *Proc. Natl. Acad. Sci. U.S.A.* **86,** 9119.

Lee, M. G.-S., and Van der Ploeg, L. H. T. (1990). *Science* **250,** 1583.

Le Page, R. W. F., Wells, J. M., Prospero, T. D., and Sternberg, J. (1988). *In* "Current Communications in Molecular Biology: Molecular Genetics of Parasitic Protozoa." (M. J. Turner and D. Arnot, eds.) p. 58. Cold Spring Harbor, New York.

Moyzis, R. K., Buckingham, J. M., Cram, L. S., Dani, M., Deaven, L. L., Jones, M. D., Meyne, J., Ratliff, R. L., and Wu, J. R. (1988). *Proc. Natl. Acad. Sci. U.S.A.* **85,** 6622.

Murray, A. W., and Szostak, J. S. (1983). *Nature (London)* **305,** 189.

Parsons, M., Nelson, R. G., and Agabian, N. (1984). *Immunol. Today* **5,** 43.

Pays, E., and Steinert, M. (1988). *Annu. Rev. Genet.* **22,** 107.

Pluta, A. F., and Zakian, V. A. (1989). *Nature (London)* **337,** 429.

Pologe, L. G., and Ravetch, J. A. (1986). *Nature (London)* **322,** 474.

Rudenko, G., Le Blancq, S., Smith, J., Lee, G.-S. M., Rattray, A., and Van der Ploeg, L. H. T. (1990). *Mol. Cell. Biol.* **10,** 3492.

Sloof, P., Bos, J. L., Konings, A. F. J. M., Menke, H. H., Borst, P., Gutteridge, W. E., and Leon, W. (1983). *J. Mol. Biol.* **167,** 1.

Sternberg, J., and Tait, A. (1990). *Trends Genet.* **6,** 317.

Ten Asbroek, A. L. M. A., Ouellette, M., and Borst, P. (1990). *Nature (London)* **348,** 174.

Tobin, J. F., Laban, A., and Wirth, D. (1991). *Proc. Natl. Acad. Sci. U.S.A.* **88,** 864–868.

Van der Ploeg, L. H. T., Valerio, D., De Lange, T., Bernards, A., Borst, P., and Grosveld, F. G. (1982). *Nucleic Acids Res.* **10,** 5905.

Van der Ploeg, L. H. T., and Cornelissen, A. W. C. A. (1984). *Philos. Trans. R. Soc. Lond. B.* **307,** 13.

Van der Ploeg, L. H. T., Schwartz, D. C., Cantor, C. R., and Borst, P. (1984a). *Cell* **37,** 77.

Van der Ploeg, L. H. T., Cornelissen, A. W. C. A., Barry, J. D., and Borst, P. (1984b). *EMBO J.* **3,** 3109.

Van der Ploeg, L. H. T., Liu, A. Y. C., and Borst, P. (1984c). *Cell* **36,** 459.

Van der Ploeg, L. H. T., Smith, C. L., Polvere, R. I., and Gottesdiener, K. (1989). *Nucleic Acids Res.* **17,** 3217.

Van der Ploeg, L. H. T. (1990). *In* "Frontiers in Molecular Biology: Genome Rearrangements and Amplification" (Hames and Glover, eds.), pp. 51. IRL Press.

Walmsley, R. W., Chan, C. S. M., Tye, B-K., and Petes, T. D. (1984). *Nature (London)* **310,** 157.

Weiden, M., Osheim, Y. N., Beyer, A. L., and Van der Ploeg, L. H. T. (1991). *Mol. Cell. Biol.* **11,** 3823–3834.

Williams, R. O., Young, R. O., and Majiwa, P. A. O. (1982). *Nature (London)* **299,** 417.

Wells, J. M., Prospero, T. D., Jenni, L., and Le Page, R. W. F. (1987). *Mol. Biochem. Parasitol* **24,** 103.

17

Use of DNA Sequence Homology and Pseudogenes for the Construction of Active Variable Surface Antigen Genes in *Trypanosoma equiperdum:* How Ordered Expression Is Established

HARVEY EISEN
Fred Hutchinson Cancer Research Center
Seattle, Washington 98104

INTRODUCTION

African trypanosomes are extracellular hemoflagellates that are responsible for widespread disease in humans and domestic animals. These parasites evade the host immune system by periodically changing their variable surface glycoproteins (VSGs), which are thought to be the unique parasite antigens presented to the host immune system. Whereas the mechanism of this antigenic variation is not yet understood, it appears that activation of the VSG genes is a complex process that involves multiple steps. Silent or basic copy (BC) VSG genes are duplicatively transposed into one of several telomere-linked expression sites where they can be transcribed as expression-linked copies (ELCs) [reviewed by Donelson and Rice-Ficht, (1)]. Since transcription of the VSG genes can be regulated within expression

Mechanisms of Eukaryotic DNA Recombination

sites, their insertion into an expression site, although necessary for transcription, is not sufficient (2–7).

EARLY- AND LATE-APPEARING VARIABLE SURFACE GLYCOPROTEINS

The different antigenic variants do not appear randomly during the course of an infection. Rather, as has been shown for *Trypanosoma equiperdum* and *T. brucei*, the variants appear at a predictable time in a loosely defined order (8–10). Furthermore, transfer of any *T. equiperdum* variant to a nonimmune host results in the reappearance of the initial variant, variant antigen type (VAT)-1. Thus, the trypanosomes express their variant antigens in an order, and have a mechanism for resetting the variant clock in a nonselective environment.

EARLY GENES HAVE COMPLETE SILENT COPIES, WHEREAS LATE SILENT COPY GENES ARE DEFECTIVE

We have shown that VSGs appearing early in a *T. equiperdum* infection are encoded by complete silent copies (11). On the other hand, the late-appearing VSGs appear to be coded for by composite genes that exist in complete form only when found in expression sites. The 3′-most 200–300 base pairs (bp) of the late-appearing ELCs are donated by one silent copy gene, the 3′ donor, whereas the remaining 5′ portion is donated by an *unrelated* silent gene or family of genes. In each case thus far examined, there is a short region (75–100 bp) of homology between the 5′ and 3′ donor genes, and it is within this region that they are joined in the ELC (12–15).

We have determined the nucleotide sequences of the silent donor genes used for the construction of seven late-appearing *T. equiperdum* VSGs (12–15). All, including 5′ and 3′ donor genes, are pseudogenes in that they do not have sufficient open reading frame (ORF) to encode the VSG. The silent pseudogenes contributing the 5′ segments of the ELCs lack functional 3′ ends. The silent genes contributing the 3′ segments are mutated in their 5′ regions. It is therefore likely that all the silent donor genes used for the construction of the late ELCs are pseudogenes. Although the four VSG-78 ELCs were all different from one another in their 5′ segments, all of their nucleotide sequences could be accounted for as being derived from different segments of the same set of donor pseudogenes. Thus, all were mosaics derived from various members of the 78 5′ donor family. This strongly suggests that the sequences of the ELCs were generated by homologous recombination of the silent pseudogenes to obtain complete (ORFs).

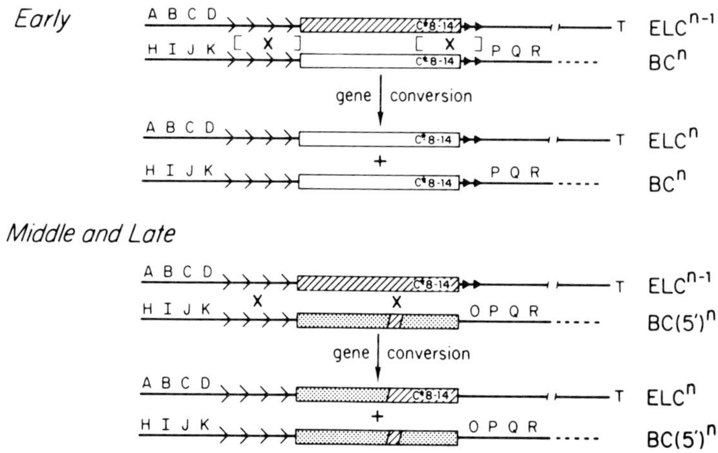

Fig. 1. Model for the order of expression of the VSGs. The flanking regions of early expressed VSG BCs (BC^n) have some sequence homology with the expression sites, allowing transposition to occur independent of the gene occupying the expression site. Early BCs also have complete open reading frames. In contrast, late-expressed BCs [$BC(5')^n$] do not encode complete VSGs, or miss the homology with the expression sites. Therefore, their expression is dependent on a recombination with the VSG gene present in the expression site at the time. Elements conserved 3′ of VSG coding regions are indicated; Cn rich in cytosine, 8: TGCTACTT, 14: TGATATATTTTAAC. Large arrows represent sequences common to the 5′ regions of BCs and expression sites; small arrows represent sequences common to the 3′ regions of BCs and expression sites; T: telomere.

THE USE OF DNA SEQUENCE HOMOLOGY TO GENERATE ORDERED EXPRESSION OF THE LATE-APPEARING VARIABLE SURFACE GLYCOPROTEIN GENES

The finding that the early-appearing VSGs are encoded by complete silent genes, whereas those appearing late are are constructed *de novo* from pesudogenes, suggested a model for antigenic variation based on homologous recombination (16). This is shown schematically in Fig. 1. Complete donor genes share homology with the expression sites outside of the open reading frame of the resident VSG gene, and therefore would have a high probability of converting an expression site containing *any* VSG gene. These silent copies would be used early in an infection. On the other hand, the multiple events required for the formation of the composite ELCs would result in low probability of use and late appearance. The 5′ donor genes of the late-appearing VSGs have very defective 3′ ends and have recombined with the 3′ donor genes within the ORFs. Since all of the events that we have characterized have occurred in regions of sequence homology, it is likely that

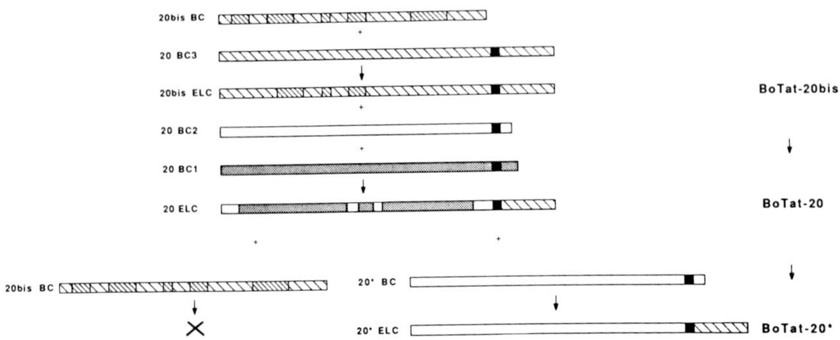

Fig. 2. This diagram illustrates how the VSG-20, or the VSG-20* 5′ donors could productively recombine within VSG-20bis ELC, or how they could replace each other in an expression site using a small region of sequence identity at their 3′ ends. VSG-20bis BC could recombine with VSG-20 BC3 within a large region of homology between the two genes, but not with VSG-20 or VSG-20* ELCs, because it does not share enough sequence similarity with them. The region of sequence similarity between VSG-20 BC2, VSG-20* BC, and VSG-20 BC3 is represented by a black box. The regions of VSG-20bis BC and VSG-20 BC that have related sequences are represented by diagonal stripes.

the process is one of homologous recombination (conversion). It was therefore proposed that defective 5′ donor genes can be recombined into an expression site only if they have homology with the resident ELC in that site. However, where homology existed, the particular silent gene would have a high probability of use. This would generate an ordered expression of the late-appearing genes.

The model described above makes the following predictions that we have tested: (1) early-appearing silent VSG genes can convert an expression site independent of which gene occupies the site; (2) late-appearing silent VSG genes can convert an expression site only if they have sequence homology with the current resident of that site; and (3) joining of the 5′ and 3′ segments of late-appearing VSG genes occurs only at the expression sites and not in "solution" in the nucleoplasm or cytoplasm.

These predictions were tested by isolating two independent variants (VSG-20* and VSG-20bis) that activated the same 3′ donor gene as the previously described VSG-20, but used different 5′ families (15). The three variants were recognized as immunologically different in mice. All of the contributing silent copies were cloned and sequenced. As previously found for the 78 and 20 genes, all were pseudogenes. The VSG-20* and VSG-20bis ELCs differed from one another and from VSG-20 in their 5′ segments. From the sequences of the silent and ELC genes, it was apparent that the VSG-20bis ELC was formed by conversion of the VSG-20 3′ donor gene in the segment 5′ to its region of homology with the VSG-20 5′ donor genes. The 5′ donor gene for VSG-20* ELC is homologous with the 3′ donor gene in the same region as are the VSG-20 5′ donor genes. However, neither the VSG-20 nor the VSG-20* 5′ donors (nor the respective ELCs) have homology with the VSG-20bis 5′ donor gene. Thus, our model for gene formation through the homologous recombination of the pseudogenes made the strong prediction that VSG-20bis could give rise to both VSG-20 and VSG-20*. Similarly, VSG-20 and -20* could interconvert. However, neither could give rise to VSG-20bis, since neither has homology with the 20bis 5′ donor. This is shown schematically in Fig. 2. Furthermore, all should give rise to the early variant VAT-1, and none should give rise to other late variants with whose 5′ donor genes they have no homology.

We used polymerase chain reaction (PCR) analysis to test our predictions. It was reasoned that if variant x can give rise to variant y, variant y will be present as a contaminant in a population of x. Specific PCR primers were prepared for each of the three VSG-20 ELCs, for VSG-1 (an early variant) and for VSG-28, another late variant. RNA was prepared from each of the variants, reverse transcribed, amplified with the appropriate primers, and the products were analyzed in agarose gels. By mixing VAT-20 and VAT-20bis before RNA extraction, it was determined that one cell of VAT-20bis could be detected in 10^9 Vat-20 cells. Similarly one VAT-2bis could be detected in 10^9 VAT-20* cells.

As predicted, VSG-1 sequences were found in all the RNA samples, and VSG-28 sequences were found only in the VAT-28 RNA. VSG-20 and -20* sequences were found in the RNAs from VATs-20, -20* and 20*. VSG-20bis sequences were found *only* in VAT-20bis RNA. Finally, VSG sequences could be amplified from all the samples using PCR primers from sequences conserved in all VSG ELC genes.

These results confirm all the predictions stated above and strongly argue that the fundamental mechanism generating the late ELCs relies on sequence homology between the resident ELC in an expression site and the "incoming" 5′ (or by extrapolation, 3′) donor genes. Other mechanisms, such as differences in growth rates of different variants, might well modulate this effect. However, these hypothetical mechanisms can function only after the formation of the VSG genes.

5′–3′ JOINING

As stated above, we have not found that the 5′ and the 3′ donor genes contributing to a given VSG ELC belong to the same gene families. Rather, they share short regions of homology within which their joining occurs.

Comparison of these sequences among donors of different VSG ELCs reveals that they are different, as expected. However, all thus far sequenced contain an 11-base pair sequence at, or very near the site of recombination. This sequence, TGCAAACCCCC, can be in either orientation. In the component genes of VSG-78, it is as written (12). However, in the components of VSG-20, it is in the opposite orientation (14). This raises the possibility that the 3′–5′ joining may actually be a site-specific recombination event. However, our results strongly argue more extensive homology specific to the components of each gene is essential. Perhaps these homologous sequences allow for pairing of the two partners so that the putative site-specific event can occur.

A POSSIBLE MOLECULAR MECHANISM FOR LATE-VARIABLE SURFACE GLYCOPROTEIN EXPRESSION-LINKED COPY FORMATION

The extent of genetic mosaicism observed in the 5′ segments of the late-appearing VSG ELCs can be very great (14 conversion patches within 750 base pairs in VSG-78). Also, where it has been possible to determine the number of genes that contributed to a given 5′ mosaic, it appears to be two. This raises the possibility that the mosaics are made via reverse transcription. It is possible that the silent genes are transcribed at very low levels, or sporadically, and that they form dimers, which are reverse transcribed with strand switching of the polymerase. This, when made double-stranded, could integrate into an expression site if it found homology with the resident gene. It is also possible that the late ELCs are generated by multiple rounds of gene conversion by the 5′ donor genes on the previous resident of the expression site. We are now investigating these possibilities.

WHY PSEUDOGENES?

If the role of antigenic variation in the life of the parasites is to escape the host immune system, then its main function is to generate diversity. The use of defective genes guarantees the generation of diversity, since it appears unlikely that recombination of the same set of pseudogenes would give the same result twice. Furthermore, since ELCs can be inactivated to become silent genes, this would amplify the rate of evolution of the silent gene repertoire. It is likely that the complete silent genes are just such recently produced silent copies. Whereas the maintenance of these complete silent copy genes does not increase the diversity, it may be important in the establishment of critical numbers of the parasites in the infected mammal. Since the infecting inoculum in the wild may be small, the pseudogene mechanism for ELC generation might be too inefficient for the parasites to survive in the infected host. However, the availability of the complete silent genes would guarantee this.

ACKNOWLEDGMENTS

I should like to thank my colleagues Genevieve Thon and Charles Roth. This work was supported by grants from the National Science Foundation and the National Institutes of Health.

REFERENCES

1. Donelson, J. E., and Rice-Ficht, A. C. (1985). *Micobiol. Rev.* **49,** 107–125.
2. Williams, R. O., Young, J. R., and Majiwa, P. A. O. (1979). *Nature (London)* **282,** 847–849.
3. Pays, E., Van Meirvenne, N., Le Ray, D., and Steinert, M. (1981). *Proc. Natl. Acad. Sci. U.S.A.* **78,** 2673–2677.
4. Young, J. A., Miller, N., Williams, R., and Turner, M. (1983). *Nature (London)* **306,** 196–198.
5. Bernards, A., De Lange, T., Michels, P., Liu, A. Huisman, M. J., and Borst, P. (1984). *Cell* **36,** 163–170.
6. Buck, G., Jacquemot, C., Baltz, T., and Eisen, H. (1984). *Gene* **32,** 329–336.
7. Buck, G., Longacre, S., Raibaud, A., Hibner, U., Giroud, C., Baltz, T., Baltz, D., and Eisen, H. (1984). *Nature (London)* **307,** 563–566.
8. Van Meirvenne, N., Janssens, P. G., and Magnus, E. (1975). *Soc. Belge Med. Trop.* **55,** 409–423.
9. Capbern, A., Giroud, C., Baltz, T., and Mattern, P. (1977). *Exp. Parasitol.* **42,** 6–13.
10. Myler, P., Allen, A., Agabian, N., and Stuart, K. (1985). *Infect. Immun.* **47,** 684–690.
11. Longacre, S., and Eisen, H. (1986). *EMBO J.* **5,** 1057–1063.
12. Roth, C., Longacre, S., Raibaud, A., Baltz, T., and Eisen, H. (1986). *EMBO J.* **5,** 1065–1070.
13. Roth, C., Bringaud, F., Layden, R. E., Baltz, T., and Eisen, H. (1989). *Proc. Natl. Acad. Sci. U.S.A.* **86,** 9375–9379.
14. Thon, G., Baltz, T., and Eisen, H. (1989). *Genes Dev.* **3,** 1247–1254.
15. Thon, G., Baltz, T., Giroud, C., and Eisen, H. (1990). *Genes Dev.* **3,** 1374–1383.
16. Eisen, H., Longacre, S., and Buck, G. (1985). *In* "The Impact of Gene Transfer Techniques in Eukaryotic Biology" (P. Starlinger, ed.) pp. 49–53. Springer-Verlag, Berlin.

18

Chromosomal Size Variations in *Plasmodium falciparum*

ALAN F. COWMAN AND DAVID J. KEMP
The Walter and Eliza Hall Institute of Medical Research
Victoria 3050, Australia

INTRODUCTION

The protozoan parasite *Plasmodium falciparum* causes the most severe form of human malaria, and it is therefore the subject of detailed study in an effort to devise an effective vaccine (1). This has resulted in major advances in our understanding of genetic diversity in this parasite. Many genes, encoding polypeptides that are antigens during natural infection, have been cloned. Surprisingly, it has been found that many of these genes encode tandem repeats of oligopeptide sequences and that a major portion of the host immune response is directed against these structures. Variation in these peptide repeats, presumably driven by the host immune system, forms a major component of our current knowledge of genetic diversity in *P. falciparum*.

The life cycle of malarial parasites includes a sexual phase in the mosquito, the vector for these plasmodia, enabling gametes of different parasites to undergo cross fertilization to exchange chromatin via meiosis. In this way parasites that have escaped destruction by the immune system of their vertebrate host or managed to develop resistance to the few antimalarials in use, can efficiently pass these traits on to a further generation of parasites. If the genetic basis for this spread of antigenic variation is not understood, then the future of any vaccine against malaria will be matter of chance, and the continuing responsiveness by the parasite to new drugs will be uncertain.

The advent of pulsed field gradient (PFG) gel electrophoresis (2) coupled

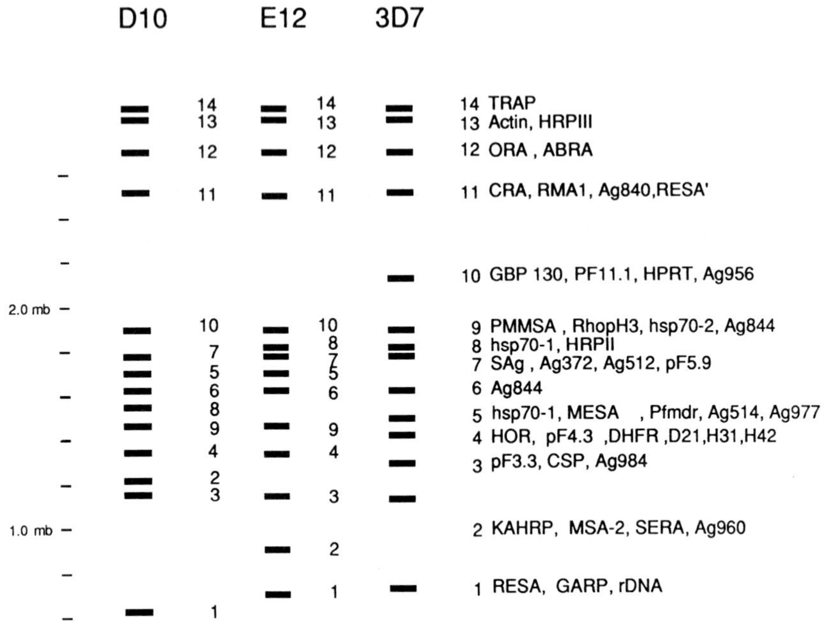

Fig. 1. Karyotype map of D10, E12 (two subclones of isolate FC27) and 3D7. A scaled representation of the 14 chromosomes and genes encoded by each. References for these cloned genes can be found in the review by Kemp *et al.* (1). The cloned line 3D7 has been taken as the standard, as it can traverse the life cycle of the parasite.

with the increasing number of cloned probes has allowed the study of the structure of *P. falciparum* chromosomes and will ultimately facilitate a more detailed account of the genetics of this organism. The genome of *P. falciparum* is organized into 14 chromosomes (Fig. 1), a figure that has been arrived at independently through the resolution of the number of chromosome bands by PFG electrophoresis (3,4) and also by counting kinetochores after reconstruction of a nucleus following electron microscopy of serial sections (5). The parasite is haploid in the asexual phase of the life cycle, and it has become clear that there is considerable variation in chromosome size from one parasite clone to another. We discuss in this chapter results that suggest a major component of this variation is the recombinational expansion, contraction, and deletion of repetitive subtelomeric sequences. A second component of chromosome size variation that has recently been identified involves amplification of the *P. falciparum* homolog of the multiple drug resistance (*mdr*) gene, a process that may have been driven by drug pressure in the field.

INTRAGENIC RECOMBINATION IN A MAJOR ANTIGEN OF *PLASMODIUM FALCIPARUM*

The precursor to the major merozoite surface antigen (PMMSA) is a 200-kilodalton protein antigen of *P. falciparum* that varies between different isolates. DNA sequence analysis of the gene from a number of different isolates (6–11) has revealed that the gene can be divided into a number of blocks that vary with the degree of conservation between different isolates. The variable blocks that have been analyzed so far occur in two different forms, and therefore the antigen shows dimorphism. At the 5′ end of the gene is a relatively short block (block 2) of tandem repeats that shows some polymorphism. In different isolates of *P. falciparum*, a number of different combinations of these blocks exist as a result of recombination across the conserved blocks (Fig. 2). These events have been extensively analyzed in a large number of isolates by synthesizing short oligonucleotides corresponding to the blocks of the dimorphic alleles and hybridizing of these to genomic DNA. Of the 23 isolates examined by either DNA sequence analysis or hybridization with oligonucleotide probes, all except CSL2 and RO-33 conform to the dimorphic allele model. These have a region in variable block 2 that is different from the others. Consequently, the dimorphic model holds for the PMMSA gene, except for block 2 that appears to be polymorphic. This raises the possibility that many other forms of this antigen exist in nature. Analysis of recombination events within the gene has shown that recombination has occurred within the conserved blocks 3 and 5 but not blocks 12 (8,9) or blocks 6 and 16 in 22 of the 23 isolates analyzed. It would appear, therefore, that recombination between some blocks is selected against, presumably because they generate nonfunctional polypeptides. Since the parasite is haploid, these recombination events must have occurred during meiosis. This recombinational reassembly of the dimorphic alleles of the PMMSA gene can explain most of the antigenic variation in this molecule.

SIZE POLYMORPHISMS IN *PLASMODIUM FALCIPARUM* CHROMOSOMES

The 14 chromosomes of *P. falciparum* vary in size from approximately 600 to 3000 kb, and homologous chromosomes in independant isolates are highly polymorphic in size (12,13). These size polymorphisms have been observed in cloned lines of *P. falciparum* and also in isolates taken directly from infected individuals under conditions that would not allow selection *in vitro*

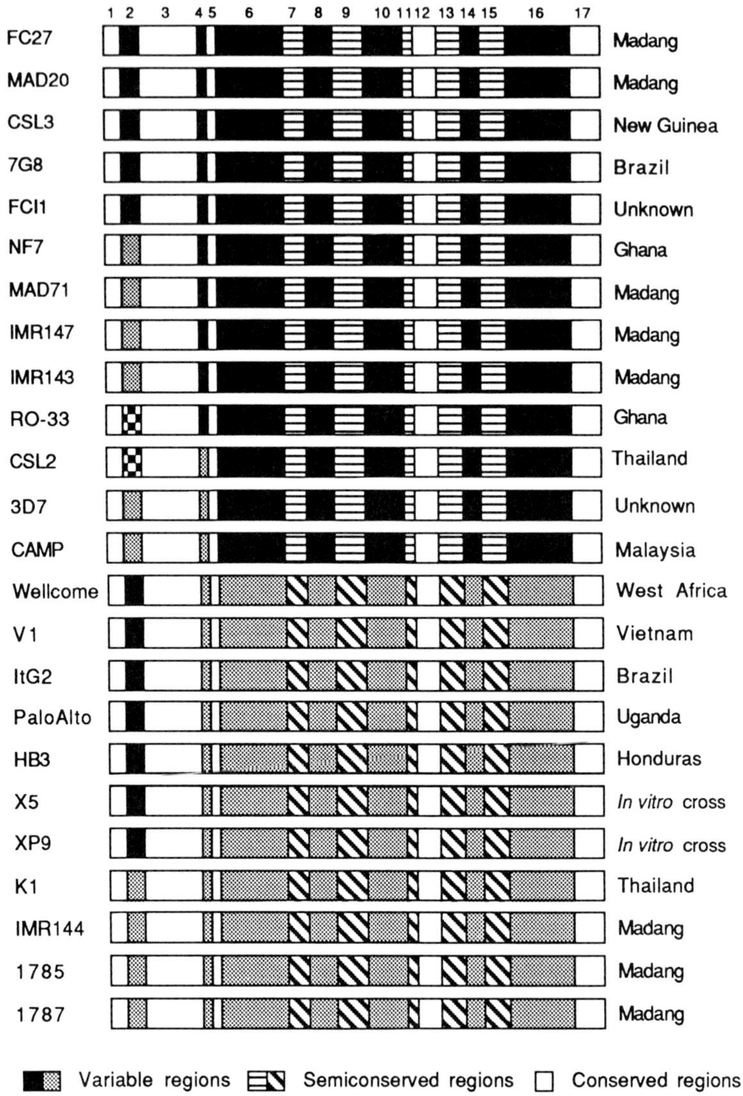

Fig. 2. Schematic of the amino acid sequences of the 23 PMMSA molecules, using the convention of Tanabe *et al.* (9). The structure of the MAD20, CAMP, and Wellcome isolates were derived from their previously published sequences (8–10). All other variable boxes, represented either by filled boxes (FC27-like) or likely shaded boxes (K1-like), have been determined using oligonucleotide probes (8). Conserved regions (open boxes) and semiconserved regions (boxes with horizontal and diagonal line) were inferred from the dimorphic model (9).

(13). Also it has been shown in a cloned line that the chromosomes were identical in all of the stages of the asexual cycle of the parasite.

Hybridization of cloned DNA molecules to blots of PFG gels has allowed the assignment of 45 genes to the 14 chromosomes (Fig. 1), and this has revealed that genetically equivalent chromosomes show marked size variation (14). For reference purposes, the isolate 3D7 was chosen for numbering the chromosomes, as it can traverse the entire life cycle. Hybridization of the probes to three cloned lines (D10, E12, and 3D7) allows unambiguous assignment of any gene to the relative chromosome. Current evidence suggests that these polymorphisms do not result from interchromosomal translocation events but are a result of intrachromosomal events, usually deletions if the changes have occurred *in vitro*.

CHROMOSOME SIZE POLYMORPHISMS GENERATED BY DELETION OF SUBTELOMERIC REPEATS

Restriction maps for chromosome 1 and 2 of six cloned lines of *P. falciparum* have been generated (Fig. 3) (15). Two of the cloned lines were D10 and E12, which differ in the size of both chromosome 1 and 2, even though they were both derived from the same parent isolate FC27. The other clones were 3D7 and HB3, the parental clones used for a genetic cross (16), and two of the progeny derived from this cross. The restriction maps, defined with seven different enzymes that cut the (AT)-rich genome rarely, delineate the locations of eight genetic markers, including genes for five antigens (Fig. 3).

In all the parasites, chromosome structure is conserved in the central regions, but is polymorphic both in length and sequence near the telomeres (15). A telomere probe, originally isolated from *Plasmodium berghei* (17), was known to hybridize to the telomeres of *P. falciparum*. The telomeric repeats are restricted to an area of less than or equal to 15 kb from the end of the chromosomes, and are associated with a microheterogeneity in fragment size that is characteristic of telomeric sequences (18). An *Apa*I site occurs frequently in a conserved position 12–15 kb from the end of the chromosome in many cases, even in nonhomologous chromosomes.

A *P. falciparum*-specific repetitive DNA element has been described (19–21) that consists of a 21-bp core sequence and has been named rep20 (15). Clusters of this repeat occur in blocks containing at least 13 repeats, which can measure up to and greater than 15 kb. Rep20 sequences have been reported to reside on all chromosomes (12); however, chromosome 1 in D10 and chromosome 2 in E12 lack detectable rep20 sequences (Fig. 3) (15). The

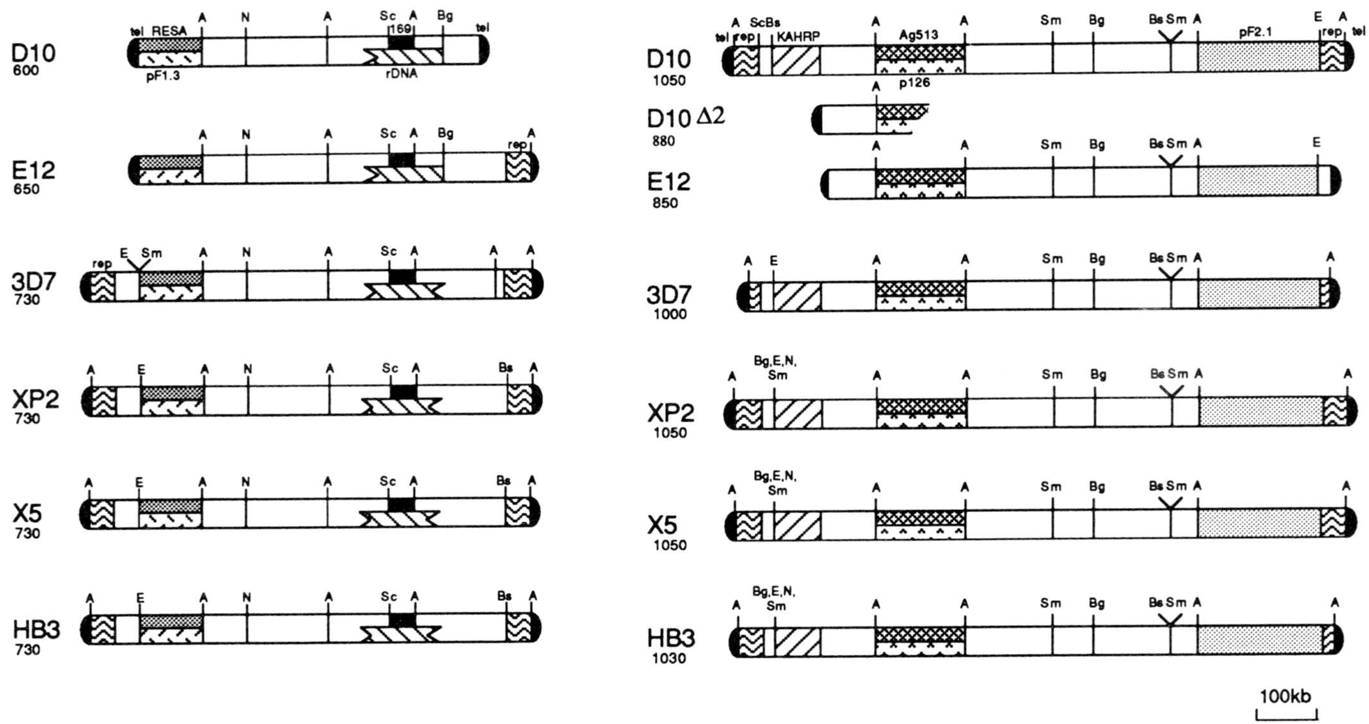

Fig. 3. Physical maps of *P. falciparum* chromosomes 1 and 2 in six cloned lines. Chromosome 1 is shown at the left and chromosome 2 at the right. The figures at the left of each chromosome indicate the sizes (in kb). The position of each sequence, RESA (ring-infected erythrocyte surface antigen), pF1.3 (an anonymous DNA *Eco*RI fragment from chromosome 2), tel (telomere), KAHRP (knob-associated histidine-rich protein), rep (rep20), rDNA (ribosomal DNA), Ag 513, p126 is shown as the smallest fragment to which a probe hybridized. Starting at the top, the first time each sequence occurs on a chromosome, it is labeled with the name of the probe. When it appears again, lower in the figure, the region is identified by the shading pattern. Black semicircles represent telomeres. Restriction enzymes used: A, *Apa*I; Bg, *Bgl*II; Bs, *Bss*HII; N, *Nar*I; Sc, *Sac*II; Sm, *Sma*I. The structure of chromosome 2 in D10 Δ2 is identical to that in D10, except in the region shown.

rep20 sequences are associated with the ends of chromosomes, just adjacent to the telomeres (15). Chromosome 1 of E12 has a rep20 cluster at only one end, and it has been proposed that the generation of chromosome length polymorphism occurs by recombination within, and deletion of, expendable rep20 repeats (15,22). The maps for chromosome 1 and 2 of the six cloned lines indicate strong structural conservation in central regions. The length polymorphisms coincide with fragments bearing rep20 repeats, and it is evident that the blocks of rep20 can vary in copy number.

Small telomeric *Apa*I fragments that included the rep20 sequences have been found on all chromosomes (15). However, a number of chromosomes from some clones have one larger telomeric *Apa*I fragment, and all of these lack one of the rep20 domains. Consequently, it appears that the lack of a small *Apa*I telomeric fragment correlates with the absence of the rep20 domain. It can be concluded, with the exception of such deletions that vary from isolate to isolate, that the general structure of each end of every chromosome is the same. It has been recently found that rep20 deletions are also present in parasites taken directly from the field (23), and one isolate examined had rep20 deletions at both ends of chromosome 2. The results with field isolates show that these deletions are not *in vitro* artifacts.

DELETIONS INVOLVING CODING SEQUENCES

In several cases, size differences between homologous chromosomes of the order of 100 kb or more are known to result in the concomitant loss of genes encoding antigens. One such gene KAHRP (knob-associated histidine-rich protein), normally located on chromosome 2, is completely (12,24,25) or partially (26) deleted from the genome. Pologe and Ravetch (26) showed that a remnant of the partially deleted KAHRP gene was located near the telomere. Also, analysis of the cloned line E12 and a D10 variant has shown that deletions have removed all rep20 sequences and the *Apa*I site usually adjacent to the KAHRP gene (15). A second gene we have termed RESA (ring-infected erythrocyte surface antigen) encodes a highly conserved antigen (27), and this protein has been found not to be expressed in the isolate FCR3 (28). Mapping of chromosome 1 from FCR3 by PFG analysis has shown that a portion of the RESA gene has been inverted and partly deleted along with a typical rep20 deletion that removes the *Apa*I site and ~ 120 kb of DNA between RESA and the adjacent telomere in FCR3 (Fig. 4a) (28). Sequence analysis of the 5′ region of the deleted RESA gene has shown that the telomeric repeat is attached to a 98-bp sequence, which is attached in turn to the rest of the RESA gene of FCR3. It has recently been shown that this 98-bp region consists of a portion of the very 5′ end of the gene, but

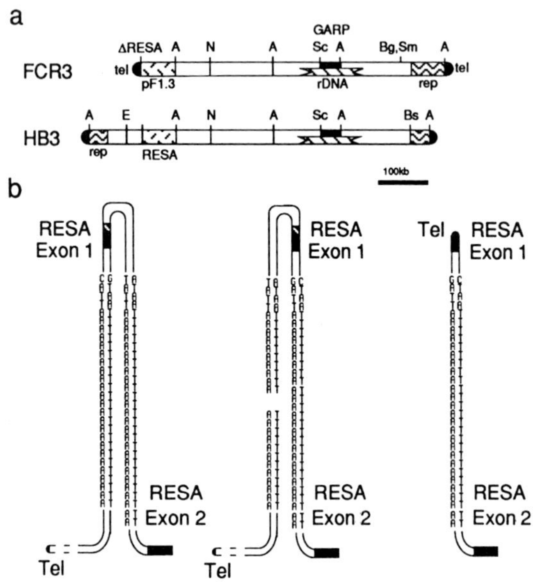

Fig. 4. (a) Physical map of chromosome 1 in FCR3 and HB3. The position of the RESA gene is indicated by shaded areas bounded by the restriction sites. A, *Apa*I; Bg, *Bgl*I; Bs, *Bss*HII; E, *Eag*I; N, *Nar*I; Sc, *Sac*II; Sm, *Sma*I. (b) Model for the deletion. At the left is shown the left end of the undeleted chromosome 1, from the telomere to exon 2 of RESA, folded to align the two homopolymer regions. In the center is shown a hypothetical recombination intermediate, where RESA exon 1 is inverted, but the telomere-proximal break point has failed to rejoin. At the right is shown the structure determined for FCR3.

inverted with respect to the rest of the gene. A model for the inversion and breakage of this chromosome can be devised because in the wild-type chromosome, the 98 bps are preceded by a long stretch of As, whereas the deletion point is followed by a region of Ts (Fig. 4b). A recombination event across these two regions and failure of the telomere proximal break to rejoin would generate the structure of chromosome 1 in FCR3. The broken end of the chromosome would then presumably be degraded until an appropriate substrate sequence for the telomere terminal transferase was uncovered. Simple breakage of the chromosome followed by telomeric addition appears to account for some other deletions of the subtelomeric region (29).

THE MULTIDRUG RESISTANCE GENE OF *PLASMODIUM FALCIPARUM* AND CHROMOSOME SIZE POLYMORPHISMS

The similarity between chloroquine resistance in *P. falciparum* and the phenomenon described in multidrug-resistant neoplastic cell lines (30,31) has suggested that the same gene may be involved in both phenotypes (32). Analyses of mouse and human tumor cell lines has shown that overexpression of the protein product (P-glycoprotein) of a gene called the *multidrug resistance* (*mdr*) gene mediates rapid efflux of a diverse range of antitumor drugs from the cell (30,31). The tumor cell lines contain amplified domains of DNA containing the *mdr* gene that results in a large increase in expression of the P-glycoprotein (33,34). This drug-resistance phenotype is reversed by verapamil, and a number of other calcium channel blockers, by inhibiting the efflux of the drugs from the cell. The same phenomenon has also been observed in multidrug-resistant *P. falciparum*, and this phenotype can also be reversed using verapamil (32).

The *P. falciparum* homolog (*pfmdr*) to the human and murine *mdr* gene has recently been isolated (35), and it has been shown that it is amplified in some chloroquine-resistant cloned lines of the parasite. There appears to be a strong link between the *mdr* gene and chloroquine resistance, but it has yet to be demonstrated conclusively that this gene or a member of the *mdr* gene family is involved in the multidrug-resistance phenotype in *P. falciparum*. Nevertheless, it is clear that amplification of the *mdr* gene and regions of DNA surrounding it can result in large chromosome size polymorphisms in some *P. falciparum* cloned lines (35). Two chloroquine-resistant isolates, ITG2 and IC1, contain the *pfmdr* gene amplified between

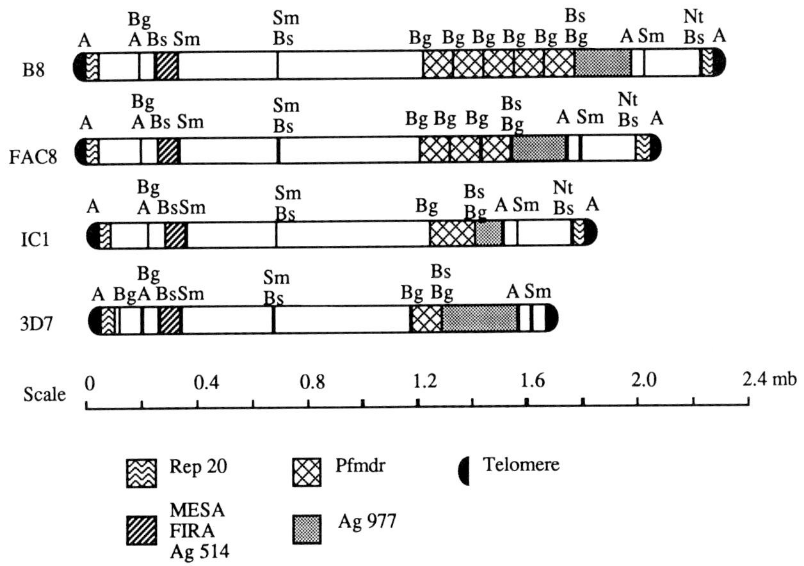

Fig. 5. Restriction maps of chromosome 5 from chloroquine-resistant ITG2 subclones (B8 and FAC8), IC1 (chloroquine resistant) and 3D7 (chloroquine sensitive). Restriction enzymes used: A, *Apa*I; Bg, *Bgl*I: Bs, *Bss*HI; Nt, *Not*I; Sm, *Sma*I. The scale is in megabase pairs.

3- and 6-fold. The *pfmdr* gene is located on chromosome 5 and, in the isolate ITG2, this chromosome is much larger and present in several different sizes as compared to other isolates. Analysis of several subclones of this parasite has shown that chromosome 5 can vary in size from clone to clone. Macrorestriction mapping of chromosome 5 from two of these clones (FAC8 and B8), another chloroquine-resistant isolate with amplification of the *pfmdr* gene (IC1) and a sensitive clone 3D7 has explained the size polymorphism (Fig. 5). From the maps shown, it can be seen that the *Bgl*I fragment, containing the *pfmdr* gene, is 105 kb in 3D7, 100 kb and present in multiple copies in FAC8 and B8, and 150 kb in IC1. Therefore the amplicon containing the *pfmdr* gene is 100 kb in FAC8 and B8 and less than 30 kb in IC1. The other difference in the size of these chromosomes is explained by a rep20 deletion from the right subtelomeric region of 3D7. Amplification of the *pfmdr* is presumably driven by drug pressure in the field; however, it is interesting that chloroquine resistance does not require amplification. This probably suggests that amplification of the *mdr* gene was not the primary event in generation of chloroquine-resistant *P. falciparum*.

CONCLUSION

It is clear that the majority of the size polymorphisms that have been identified in chromosomes of *P. falciparum* are owing to deletions of the rep20 repeat sequence in the subtelomeric regions. These deletions could occur during either mitosis or meiosis. Interestingly, many of the antigen genes encoding polymorphic proteins of *P. falciparum* are located near the subtelomeric regions of the chromosomes. Most of these genes are composed of large stretches of polymorphic repeat sequences and, in the case of PMMSA (36), variation has arisen through intragenic recombination between several different alleles. As the parasite is a haploid organism, this process could take place only during meiosis. It is conceivable that the location of these polymorphic antigen genes may increase the development of new antigenic types as they are encoded in the most dynamic region near recombining sequences of the telomeres. This may be analogous to antigenic variation in *Trypanosoma brucei*, where transcriptionally silent variant surface antigen genes are moved to an expression-linked site at the telomeres of chromosomes (37). However, a consequence of the positioning of *P. falciparum* antigen genes in these regions is that they are much more likely to be deleted because they are present in the subtelomeric regions beside the rep20 sequences. A number of examples have been described where this has occurred *in vitro* (26,28); however, it is not known whether these genes are required for growth of the parasite in the human host. It seems likely, however, that the deletion of these genes is an *in vitro* artifact, because all selection for maintenance of these antigen genes has been removed.

The selection of drug-resistant populations is accompanied by changes in chromosomal structure in a number of different organisms (38). We have described some evidence that implicates the *P. falciparum* homolog of the *mdr* gene in the development of resistance to chloroquine. Amplification of this gene in some chloroquine-resistant cloned lines results in a large increase in the size of chromosome 5 and adds a second component to the observed chromosome size polymorphisms of this parasite.

ACKNOWLEDGMENTS

The work described was supported by the Australian National Health and Medical Research Council, the John D. and Catherine T. MacArthur Foundation. A. F. C. is a Wellcome Australian Senior Research Fellow.

REFERENCES

1. Kemp, D. J., Coppel, R. L., and Anders, R. F. (1987). *Ann. Rev. Microbiol.* **41,** 181–208.
2. Schwartz, D. C. and Cantor, C. R. (1984). *Cell* **37,** 67–75.
3. Kemp, D. J., Corcoran, L. M., Coppel, R. L., Stahl, H. D., Bianco, A. E., Brown, G. V., and Anders, R. F. (1985). *Nature (London)* **315,** 247–350.
4. Wellems, T. E., Walliker, D., Smith, C. L., de Rosario, V. E., Maloy, W. L., Howard, R. J., Carter, R., and McCutchan, T. F. (1987). *Cell* **49,** 633–642.
5. Prensier, G., and Slomianny, C. H. (1986). *J. Parasitol.* **72,** 731–736.
6. Holder, A. A., Lockyer, M. J., Odink, K. G., Sandhu, J. S., Riveros-Moreno, V., Nicholls, S. C., Hillman, Y., Davey, L. S., Tizard, M. L. V., Schwarz, R. T., and Freeman, R. (1985). *Nature (London)* **317,** 270–273.
7. Mackay, M., Goman M., Bone, N., Hyde, J. E., Scaife, J., Certa, U., Stunnenberg, H., and Bujard, H. (1985). *EMBO J.* **4,** 3823–3829.
8. Peterson, M. G., Coppel, R. L., McIntyre, P., Langford, C. J., Woodrow, G., Brown, G. V., Anders, R. F., and Kemp, D. J. (1988). *Mol. Biochem. Parasitol.* **27,** 291–302.
9. Tanabe, K., Mackay, M., Goman, M., and Scaife, J. G. (1987). *J. Mol. Biol.* **195,** 273–287.
10. Weber, J. L., Leininger, W. M., and Lyon, A. (1986). *Nucleic Acids Res.* **14,** 3311–3323.
11. Certa, U., Rotmann, D., Matile, H., and Reber-Liske, R. F. (1987). *EMBO J.* **6,** 4137–4142.
12. Van der Ploeg, L. H. T., Smits, M., Ponnudurai, T., Vermuelen, A., Meuwissen, J. H. E. Th., and Langlsey, G. (1985). *Science* **229,** 658–661.
13. Corcoran, L. M., Forsyth, K. P., Bianco, A. E., Brown, G. V., and Kemp, D. J. (1986). *Cell* **44,** 87–95.
14. Kemp, D. J., Thompson, J. K., Walliker, D., and Corcoran, L. M. (1987). *Proc. Natl. Acad. Sci. U.S.A.* **84,** 7672–7676.
15. Corcoran, L. M., Thompson, J. K., Walliker, D., and Kemp, D. J. (1988). *Cell* **53,** 807–813.

16. Walliker, D., Quakyi, I. A., Wellems, T. E., McCutchan, T. F., Szarfman, A., London, W. T., Corcoran, L. M., Burkot, T. R., and Carter, R. (1987). *Science* **236,** 1661–1666.
17. Ponzi, M., Pace, T., Dore, E., and Frontali, C. (1985). *EMBO J.* **4,** 2991–2995.
18. Blackburn, E. (1984). *Cell* **37,** 7–8.
19. Bhasin, V. K., Clayton, K. C., Trager, W., and Cross, G. A. M. (1985). *Mol. Biol. Parasitol.* **15,** 149–158.
20. Guntaka, R. V., Gowda, S., Rao, A. S., and Green, T. J. (1985). *Nucleic Acids Res.* **13,** 1965–1975.
21. Zolg, J. W., Andrade, L. E., and Scott, E. D. (1987). *Mol. Biol. Parasitol.* **22,** 145–151.
22. Patarapotikul, J., and Langsley, G. (1988). *Nucleic Acids Res.* **16,** 4331–4340.
23. Biggs, B. A., Kemp, D. J., and Brown, G. V. (1989). *Proc. Natl. Acad. Sci. U.S.A.* **86,** 2428–2432.
24. Culvenor, J. G., Langford, C. J., Crewther, P. E., Saint, R. B., Coppel, R. L., Kemp, D. J., Anders, R. F., and Brown, G. V. (1986). *Exp. Parasitol.* **63,** 58–67.
25. Ellis, J., Irving, D. O., Wellems, T. E., Howard, R. J., and Cross, G. A. M. (1987). *Mol. Biochem. Parasitol.* **26,** 203–214.
26. Prologe, L. G., and Ravetch, J. V. (1986). *Nature (London)* **322,** 474–477.
27. Cowman, A. F., Coppel, R. L., Saint, R. B., Favaloro, J., Crewther, P. E., Stahl, H. D., Bianco, A. E., Brown, G. V., Anders, R. F., and Kemp, D. J. (1984). *Mol. Biol. Med.* **2,** 207–221.
28. Cappai, R., Van Schravendijk, M-R., Anders, R. F., Peterson, M. G., Thomas, L. M., Cowman, A. F., and Kemp, D. J. (1989). *Mol. Cell. Biol.* **9,** 3584–3587.
29. Pologe, L. G., and Ravetch, J. V. (1988). *Cell* **55,** 869–874.
30. Fojo, A., Akiyama, S., Gottesman, M., and Pastan, I. (1985). *Cancer Res.* **45,** 3002–3007.
31. Rogan, A. M., Hamilton, T. C., Young, R. C., Klecker, R. W., Jr., and Ozols, R. F. (1984). *Science* **224,** 994–996.
32. Martin, S. K., Oduola, A. M., and Milhous, W. K. (1987). *Science* **235,** 899–901.
33. Roninson, I. B., Chin, J. E., Choi, K., Gros, P., Housman, D. E., Fojo, A., Shen, D-W., Gottesman, M. M., and Pastan, I. (1986). *Proc. Natl. Acad. Sci. U.S.A.* **83,** 4538–4542.
34. Gros, P., Ben Neriah, Y., Croop, J. M., and Housman, D. E. (1986). *Nature (London)* **323,** 728–731.
35. Foote, S. J., Thompson, J. K., Cowman, A. F., and Kemp, D. J. (1989). *Cell.* **59,** 921–930.
36. Peterson, M. G., Coppel, R. L., McIntyre, P., Langford, C. J., Woodrow, G., Brown, G. V., Anders, R. F., and Kemp, D. J. (1988). *Mol. Biochem. Parasitol.* **27,** 291–302.
37. Van der Ploeg, L. H. T., Schwartz, D. C., Cantor, C. R., and Borst, P. (1984). *Cell* **37,** 77–84.
38. Schimke, R. T., Sherwood, S. W., Hill, A. B., and Johnston, R. N. (1986). *Proc. Natl. Acad. Sci. U.S.A.* **83,** 2157–2161.

Index

D

E

N

O

P

U

V

X

Y

Z